THE DESIGN OF ELECTRONIC EQUIPMENT

THE DESIGN OF ELECTRONIC EQUIPMENT

A Manual for Production and Manufacturing

GERSHON J. WHEELER

Prentice-Hall, Inc., Englewood Cliffs, New Jersey

Prentice-Hall International, Inc., *London*
Prentice-Hall of Australia, Pty. Ltd., *Sydney*
Prentice-Hall of Canada, Ltd., *Toronto*
Prentice-Hall of India Private Limited, *New Delhi*
Prentice-Hall of Japan, Inc., *Tokyo*

10 9 8 7 6 5 4 3

ISBN 0–13–200105–5

Library of Congress Catalog Card Number 72–172280
Printed in the United States of America

PREFACE

The designer of electronic equipment needs to know much more than he learned in school. The required information is available in bits and pieces, but since it doesn't fit any one discipline, it is not collected in one volume or set of volumes. While working for Sylvania, I prepared "the Design Guidelines Manual," a booklet which supplied the designers with basic information on manufacturing processes and other useful design data. It was enthusiastically received, and both designers and production workers suggested additional topics which might be covered.

The present volume is an enlarged and more detailed guidebook, covering material in the Sylvania manual and the additional suggestions in greater depth. Much of the information is general, suggesting *approaches* to problems rather than specific solutions. However, whenever possible, accurate quantitative and dimensional data are supplied, as well as comparative analyses of materials and techniques. The reasoning process is emphasized, rather than a blind acceptance of textbook solutions.

I would like to express my thanks to the Aluminum Association and to Chase Metals Service Division for information on metals, and to the many organizations that supplied photographs for the book. Special thanks to Sylvania for supplying excellent photographs and for having me write the original design manual.

G. J. W.

PREFACE

CONTENTS

THE DESIGN OF ELECTRONIC EQUIPMENT

1

DESIGN OBJECTIVES

Designing electronic equipment for production is not an exact science that can be learned from theory or study. In fact, if a product were given to ten different companies to produce, their designers might produce ten different designs, all meeting the specifications or requirements of the product. Furthermore, it would be a difficult exercise to choose the best of the ten solutions, for what might be best for one company could very well be the worst for another because of special proprietary techniques or equipment. Nevertheless, there are some general guidelines which can assist the designer in focusing his attention on the problem at hand and prevent him from arriving at a design which is elegant but impossible to achieve. These guidelines are the results of experience. In this chapter some of the general principles will be presented, but the designer must remember that new experiences will produce new guidelines which should be added to the growing body of knowledge.

1-1 THE MARKETS FOR ELECTRONICS

In the ideal situation a designer is called in when the equipment is being conceived. He is then in a position to assist the system engineers and electronic engineers in choosing a configuration which will be most easily and most economically produced. In most cases, unfortunately for the designer, by the time he is called upon to work on a design the electrical work is done, and in many cases the overall size and weight are fixed. In such a case, it may be comforting to point out how much easier the job would have been if only the designer had been assigned to the task earlier, but this doesn't get the job done. Although it may seem difficult, the designer must realize that the end objectives are the same as they would be if he had been called in earlier, and he should expend his efforts and time striving for those objectives.

To a large extent the market for which electronic equipment is aimed affects the specifications placed on the equipment and consequently affects the design objectives also. Thus cost is a very important criterion for a transistor radio aimed at the teenage mass market, but quality is more important than price for a hi-fi stereo amplifier. Reliability and life are major factors in designing equipment for military installations, but these seem to be less important than cabinet design in a TV-radio console.

There are two broad market areas for electronic equipment: government and civilian. In terms of dollars the government market is the larger, with the Department of Defense (DOD) being the largest single purchaser of electronic equipment in this market. Much of the money spent by DOD is for research, and although this research is slanted toward military applications, many developments do find their way into civilian use. For example, teflon-coated frying pans are a result of research in dielectric materials, and miniature hearing aids are possible because of government sponsored research in solid-state electronics. The second largest government purchaser of electronic equipment is NASA for space electronics. Again, there is a fallout of civilian-oriented devices from space programs, as well as from military electronics. DOD and NASA are by far the largest purchasers of electronic equipment, and there are many other government agencies which also use large quantities of electronic equipment. The HF receiver shown in Fig. 1-1 is an example of electronic equipment built for DOD. This receiver must

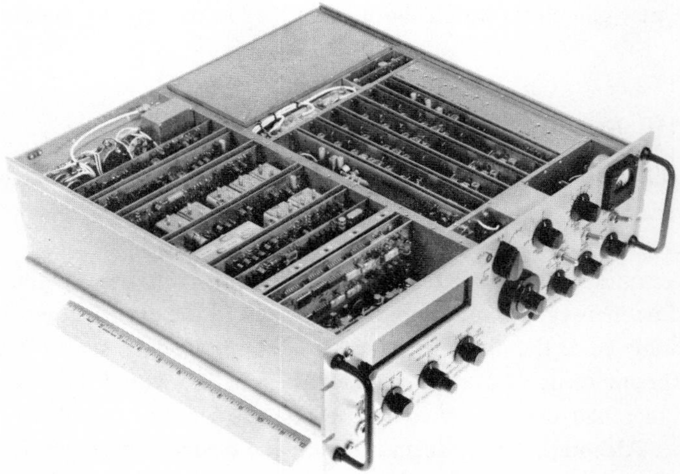

FIGURE 1-1 Military Receiver (Courtesy of GTE Sylvania Incorporated)

meet stringent operating and environmental specifications, and although cost is a consideration, it is not as important as reliability and performance.

The civilian electronics market comprises many different markets. The largest is consumer products mostly for entertainment, such as radios, television sets, and record or tape units. Figure 1-2 is a photograph of a stereo tape deck for the home entertainment market. Communication equipment and data processing represent the two largest industrial markets. A disc-operated minicomputer system is shown in Fig. 1-3. Data processing equipment such as this may be aimed at both government and civilian markets; consequently reliability and performance are important, but so is cost. Lesser civilian markets include medical electronics, industrial uses such as welding and control, and scientific or educational equipment.

These markets are continually changing as new products are developed and old ones become obsolete. Television almost eliminated radio, but automobile sets and portables enabled radio to made a strong comeback. A new development in television is the home set which accepts prerecorded cartridges and has the capability of putting an incoming program on tape

FIGURE 1-2 Stereo Tape Deck (Courtesy of Ampex Corporation)

FIGURE 1-3 Mini-computer System (Courtesy of Hewlett-Packard)

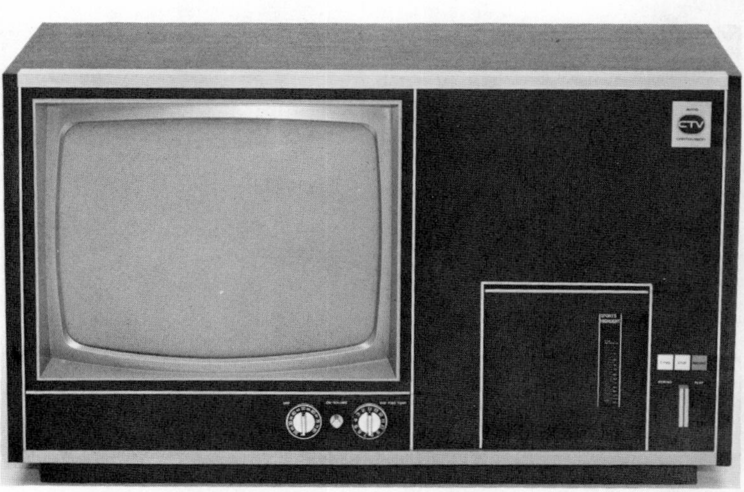

FIGURE 1-4 Cartrivision (Courtesy of Avco Corporation)

for future viewing. The "Cartrivision" set shown in Fig. 1-4 can be used to receive TV programs, as an ordinary TV set does, and can record these programs on tape if the viewer wants to save them for viewing later. A possible market in pre-taped movies or athletic events may develop. An added

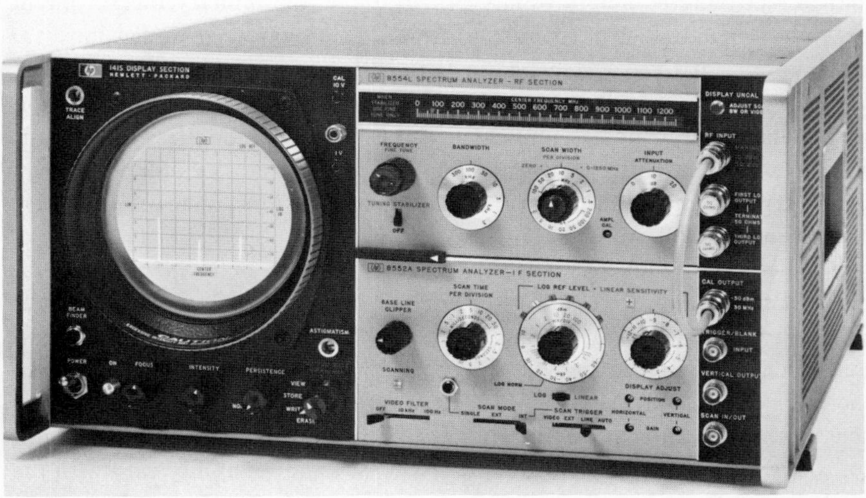

FIGURE 1-5 Spectrum Analyzer (Courtesy of Hewlett-Packard)

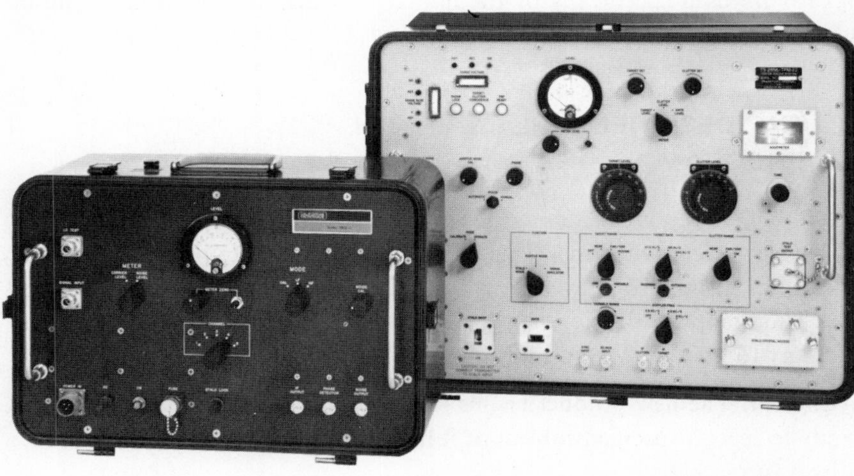

FIGURE 1-6 Military Test Equipment (Courtesy of CALIFORNIA MICROWAVE)

feature of the Cartrivision set is a TV camera attachment which permits a user to display "live" action on the screen.

These electronic equipment markets obviously have different requirements, and the designer must be aware of the specifications of the equipment and where it will be used. However, differences are not always clearly delineated, and in fact there is much overlap. For example, the military market might be characterized as one where performance and physical dimensions are more important than cost, while in the civilian market cost is paramount. Cost may be an important consideration in a hearing aid aimed at the consumer market, but not as important as performance and size. Also, cost-conscious accountants in the Pentagon are influencing the military to buy the most economical equipment that meets specifications, instead of the best equipment regardless of cost.

Whether a product is aimed at the government or civilian market, it must be tested and evaluated before release by the manufacturer, and later in operation and during routine maintenance. This requires electronic test equipment which represents another type of market of interest to all manufacturers of electronic gear. Test equipment may run from a simple inexpensive meter used to measure line voltage to a sophisticated and expensive test set to measure radar set performance. Accuracy and reliability vary with the application. The high-speed spectrum analyzer shown in Fig. 1-5 is an example of a more sophisticated test instrument which may be required for testing both government and civilian equipment. The test sets shown in Fig. 1-6 are used specifically by the military for maintenance and testing of radar equipment.

1-2 THE DESIGNER'S PLACE IN THE PROGRAM

Designing for production is one step in a rather complex development from concept to production. This one step has an impact on virtually every other stage in the program, so it is important to understand the whole process. The complete development program may vary only slightly from one company to another. A typical program follows.

1. *Concept*. The new product begins with an idea. It may be born of necessity to meet a specific problem or it may be something which could create its own market.
2. *Product study*. Engineers determine whether the product is feasible and

if so, what exactly it must do. This is an attempt to specify the mechanical and electrical requirements. If a designer is part of the product team at this stage, his suggestions and criticisms will affect the *specifications.*

3. *Preliminary design.* At first this is simply the output of the study phase; however, mechanical and electrical research and development may be necessary to prove that the specifications are realistic and to determine electrical parameters and physical dimensions. Alternatively, development tests may prove the specifications unrealistic and may necessitate further product study. However, this rarely happens if a good designer is part of the study group originally.
4. *Electrical design.* The electronics engineers begin with the preliminary design and produce a working *breadboard model.* During this phase of the process they make electrical tests on the circuits in the product and, as a result of these tests, usually modify the electrical design continually, frequently improving the electrical specifications. This can become a never-ending process, since it is apparently possible to improve the electrical performance indefinitely. Thus it may be necessary for the program manager to halt further improvements when the electrical design is satisfactory.
5. *Mechanical design.* This develops concurrently with the electrical design. The designer makes necessary mechanical studies and tests to produce a *mock-up* of the final product. As with the electrical design, results of mechanical tests can result in modifications of the mechanical design. The designer and electronics engineer should work closely together during the electrical and mechanical design stages, since the two designs are interdependent.
6. *Production design.* The breadboard model and mock-up are combined to build a *working model,* which is tested and evaluated. Electrical tests may cause further modifications of the electrical specifications, and mechanical and environmental tests may affect the mechanical specifications. As parts of the product or system are found to meet specifications or are modified to do so, design drawings are prepared. When all development is complete, a *prototype* is built. This is the *pre-production model.*
7. *Prototype evaluation.* The preproduction prototype will be tested and, if all goes well, will meet all electrical, mechanical, and environmental specifications. This rarely happens. Usually there are changes of layout or components. If the prototype fails some mechanical or environmental tests, changes in dimensions or basic materials may be made. In general,

however, if the earlier work was well done, there will be no major changes; and with slight modifications, the prototype is just that, an example of the units which will come off the production line.

It should be noted that although the pre-production prototype, as finally modified, looks exactly like the equipment which will be coming out of production, in general it is not constructed in the manner in which production units will later be made. For example, parts for this prototype may be machined on a lathe or milling machine, since only one or two prototypes are required. In production it may be cheaper to use castings or numerically controlled machines to do the same job.

The designer is the key man in the fifth and sixth steps in the above list. However, his influence is felt in every step of the program. Conversely, the results of all tests performed and all concepts promulgated in other steps affect the designer's decisions. Consequently, the designer and even the production manager should participate in the program effort from the start, especially in large programs.

1-3 GENERAL PRINCIPLES

The first and most important point for a designer to consider is how a part will be made in production. This may seem strange since the designer does not specify the method to be used in manufacturing. Nevertheless, the design often dictates the manufacturing method, so it is important for the designer to be familiar with the operation of all production machines and with all production techniques. As a general rule, he should try to choose a design which can be fabricated with tools in his company's shops or with standard tools available in any well-equipped machine shop. Special tools are costly, and there may be considerable time delay in acquiring them. If special dies or other special tools will truly result in a more economical design, a design which requires them may be justified. The important point for the designer then is that there is no fixed set of rules which specifies what is permissible and what is not. As far as tools are concerned, the designer must not take the position that standard tools are a necessity, but rather he must realize that standard tools are a desirable goal, but the extra time and cost for special tools may be justified. If he does select a design requiring special tools, he must be able to justify his selection with hard facts.

The time required to construct and assemble the finished equipment in production should be minimized, but not at a sacrifice of economy or quality. This means minimizing the number of hand operations and machine setups.

For example, it is usually faster and cheaper to set up once on a machine which will perform ten sequential operations automatically, than to require only three operations but each on a separate machine. Again it is important for a designer to know the capabilities and limitations of the tools and machines in the shop and production line.

A design must be feasible. Although designers rarely come up with a design which is elegant but impossible to assemble (it has happened), it is not infrequent that a design includes parts that are too tightly packed. The equipment is miniaturized, and this may be an advantage, but assemblers must work carefully and *slowly* to put the gear together. It is important then to leave room in the design for an assembler's hands or necessary tools to reach the assembly points as required. Also, safety during fabrication is important. There should be no sharp edges or projections which might cause injury to a worker's hands or clothes during assembly or manufacture. If extra care in handling is required, it means extra time to finish the job; this means higher cost. If a design must have fragile parts or explosive devices that require cautious handling, these elements should be installed as late in the fabrication process as possible, so that other work can be done quickly without being slowed by added safety precautions.

Reliability should be designed into the equipment. This is a more important requirement for some markets than for others, but is it a desirable objective for all markets. The surest way of designing reliability into equipment is to use a design which has been proved successful. It is important therefore to maintain a library of successful design practices and to use this library whenever possible. If a detail in the design is covered by an existing design practice documented in this library, the designer should use it. This frees the design engineer to do creative thinking on problems which have yet to be solved and also prevents him from reinventing a solution which already exists. The library of *standard practices* is an important part of a company's designing capability. With careful documentation of design successes, new engineers can be trained quickly to use the standards library for most design details and to save their creative thinking for new and unsolved problems.

This does not mean that an engineer or designer should avoid being creative when tackling a routine task. New developments or discoveries do render some designs obsolete, and engineers must be aware of changes and the necessity of updating their standard practices. It does mean that a designer's first approach to a problem should be the one in the standards library, and he should accept this approach as long as it meets the requirements of a particular job.

Ease of operation is an important design requirement. All controls which an operator must use must be within comfortable reach, and all indicators must be located where he can read them without confusion. The ideal location for controls and indicators is on the front panel of the equipment.

Since an operator should know at all times how well equipment is operating, signal lights, meters, or other function indicators should be included to display this information. It is essential to detect failure or degradation of operation or equipment. Consequently, it should be possible for the operator to notice a *change* in condition almost instantly. Instruments should be labeled in terms of what is being measured rather than by their own name. For example, a tachometer should be labeled "rpm" rather than "tachometer."

Controls should be arranged serially in order of operation. A control which affects a display, such as a curve on an oscilloscope or a meter reading, should be oriented so the movement of the control is in the same direction as the movement on the display. Controls should be far enough apart so they will not be moved accidentally when other controls are moved. Each control should be labeled, and the label should be legible with the control set in any position. The number of different controls should be kept to a minimum.

Safety is important during operation and maintenance. There must be no danger from electric shock or from exposure to harmful radiation. Heat generated in the equipment must pose no threat of burns to an operator or a maintenance man. Sharp edges, corners, and protrusions which may cause a laceration or tear clothing should be avoided.

Accessibility is the important consideration in maintenance testing and repair. The displays on the front panel, which the operator observes, are built-in test instruments since they give the first indication of a fault. Additional test points may be needed, and these should be easily accessible with standard test probes. *In no case should a test probe be located near a point of high voltage or near a point which may have a high voltage if another part fails.*

Parts which may have to be removed or replaced should be removable without disturbing other parts. There must be room for a soldering iron and other tools required to perform the repair. Parts should be labeled with their electrical values, and all labels should be clear and legible without removing other parts to get to them.

A summary of some useful general principles follows:

1. Use standard parts and standard approaches wherever possible.
2. Avoid complicated or intricate shapes or assemblies.
3. Use standard tolerances. Close tolerances cost money.

4. Minimize the number of different tools and the number of production machine set-ups required.
5. Use as few center lines as possible. Use fewest possible sizes of holes.
6. Consider safety and ease of handling during assembly and ease of operation of the finished product.
7. Arrange parts for ease of accessibility, visibility, and minimum interference with test points and other parts.
8. Use duplicate parts or subassemblies wherever possible.
9. Keep in mind how the product will be made.
10. Keep in mind how the equipment will be operated.

1-4 DESIGN GOALS

The designer's objective is to create a design which will permit *production* of the desired electronic equipment, satisfying the specific requirements of the market for which the equipment is intended. All his efforts must be aimed in this direction. He selects electrical components and mechanical devices to meet the specifications. He arranges electrical circuits and mechanical assemblies in a suitable package to comply with physical requirements, such as size and weight, environmental conditions, and location of connectors for interfacing with other equipment. Again, the market affects his selections and arrangements, and he must weigh the importance of cost, performance, reliability, and other factors as they apply to the specific user of the equipment.

The designer must think of *people* and their relationship to the equipment. During fabrication and assembly people are involved. The design must permit them to do their jobs with *convenience* and *safety*. Convenience and safety during normal operation of the equipment, as well as during servicing and maintenance, are also important considerations.

An ideal designer is a man of many talents and varied background. A prerequisite is an intimate knowledge of mechanical engineering; a working knowledge of electronic engineering is also highly desirable. He should be well-read and have a good memory, and he must keep up with latest developments in his field. If he has worked in a machine shop, that is a plus factor, since knowledge of shop techniques is most important. Designing cannot be done in isolation, since in his work a designer meets and interacts with many people. Thus he should have a genuine liking for and understanding of others. He should be creative. Above all, there is no substitute for experience.

2

STANDARDS

Standards are guidelines for all aspects of the design and construction of electronic equipment. They include techniques as well as standards of quality and workmanship. Standards cover selection of materials, choice of hardware, types of finishes, documentation practices, and materials maintained in stock.

Professional and industrial societies have set standards on equipment for civilian use, and agencies of the United States government have set standards on equipment for federal use. A partial list of sources of industrial standards includes the Institute of Electrical and Electronic Engineers (IEEE), American Society of Mechanical Engineers (ASME), Electronic Industries Association (EIA), and Underwriters' Laboratory (UL). Sources of federal standards include Armed Services Electro-Standards Agency (ASESA), National Bureau of Standards (NBS), and many others. Federal Standards, including the "MIL specifications" necessary for DOD contracts, are available from the Government Printing Office.

All industry and government standards, as well as in-house standard engineering practices, are contained in a company's standards library or standards manual. However, these are not firm commitments, but only guidelines which indicate desirable methods. Designers and engineers may occasionally take a different approach when it seems reasonable to do so. In addition, standards do become obsolete and must be replaced or updated periodically.

2-1 STANDARD PRACTICES

As mentioned in Chapter 1, an easy way to build reliability into equipment is to use methods and materials which have been used sucessfully in the past. In addition, even though the manufacturer knows that his product is reliable, it is necessary to convince a customer that this is so. When well-known and accepted methods and materials are used, the customer will be aware of the built-in reliability; thus applicable industry standards or, in cases of government contracts federal standards, should be used wherever possible. Besides the obvious selling point, a plus value as mentioned in Chapter 1 is that use of standard solutions to standard problems frees the engineer and the designer for more concentrated effort on non-standard problems.

After considering a standard solution, if an engineer or designer prefers to use a different approach, he must be able to show how and why his method is better than the standard technique. He should justify his approach *before* work is done on it, for despite the designer's confidence in his own work, the new approach may not be approved. Consequently, he must discuss his design with the program manager. There are three possible results of this discussion. First, the manager may refuse to consider the new approach, in which case the standard method must be used. Second, the new approach may be a true improvement which is then adopted as a standard, replacing the old method. The third and most common result is approval of the new method for one specific job only because of peculiarities which make the new approach better than the old one in this application but not in general.

Besides the usual electrical and mechanical standards, standard engineering practices cover such things as spacing between rivets, bolts, or other fasteners; layout; and choice of materials and finishes. It is possible that a company's own standards of workmanship may at times be more stringent than the accepted industry standards in order to offer a quality product.

The design engineer should strive for a finished product which can be built with tools found in the average well-equipped shop. Nevertheless, at times it will simplify the design if special tools such as dies or punches are made to order to perform specific tasks. These tools must then be called out on the part drawing, and this is an exception to the general rule that the equipment designer does not indicate how a part is to be made.

2-2 PREFERRED STOCK

When a designer has to select a material for a chassis, he is faced with making a choice from among hundreds of different alloys. If he decides on aluminum because of weight considerations, he must still choose from dozens of commercial alloys of this metal. Small amounts of other metals, such as copper, zinc, chrome, etc., are added to aluminum in order to accentuate a particularly desirable property. Thus, one alloy may be easily machinable, while another is more durable, and a third resists oxidation better. When the selection is made, the metal is purchased and placed in the stockroom. When the job is finished, there is always some material left in the stockroom. Now if designers are free to choose any alloy they wish, it is unlikely that two would choose the same one, since many have similar properties. Following the procedures above, after the first job was finished, the second designer would probably not select the same alloy, and at the end of the second job, there would be excess stock of two aluminum alloys in the stockroom. If the second designer knew of the first job and was willing to specify the same alloy, there would be considerable saving in both time and material. In fact, if all designers would specify one particular alloy for a chassis, the stockroom clerk could maintain this alloy in stock, knowing it would be used. This would then become a *standard stock* or *preferred stock* for building chassis. The same practice applies to other materials, hardware, and even components.

There are many advantages to having a preferred material for specific applications. First and foremost, it frees a designer from the the time-consuming job of evaluating and choosing from among dozens of materials with similar properties. Second, the material can be stocked in quantity since it will certainly be selected. Thus, it is available immediately when it is needed, and the quantity stock purchase may lower the unit price. A further advantage is that waste in the form of unused or unwanted materials left in stock is reduced, and, in the production shop, workers are less apt to use the wrong material by mistake.

Just as agreement on type of material can save time and reduce waste, so too standardization of thicknesses to be stocked will do likewise. Metal sheets 0.025 in. thick are manufactured but are not as readily available as 0.020 in. or 0.032 in. sheets. The more easily obtainable sheets should be stocked, and these sizes be *preferred sizes*.

The same considerations apply to hardware. For example, the difference in diameter between a #5 screw and a #6 screw is only 0.013 in. In general, for mechanical reasons either could be used in a particular design. Besides the screw size, the designer must specify coarse or fine thread, kind of material (brass, stainless steel, aluminum, etc.) and which of half a dozen heads to use. However, if the stockroom maintains only a few preferred sizes, and these are listed as standards to be used, the selection of screws and other hardware is greatly simplified.

The use of standard stock, including preferred materials and preferred sizes, is really a necessity. Preferred stocks include all components, hardware, and finishes which are used or may be used in more than one job. In many companies, there are even preferred stocks of packing and packaging materials.

2-3 STANDARDS MANUALS

All standard engineering practices used by a company should be documented in a Standards Manual or in its standards library. In addition, standard or preferred stock maintained in the company stockroom should be listed in a Standard Stock Catalog, which should be part of the standards library. Every designer should have ready access to these books on standards.

The company's Standards Manuals should be made a part of every contract. This is especially true in dealing with DOD and other government agencies where the tendency is to spell out in fine detail just how everything is to be done. If a company is able to offer a Standards Manual and to include its provisions in the contract, there is a big saving in time and energy during negotiation. Moreover, it is unlikely that the company will be surprised by any unusual requirements during performance of the contract.

3

METALS

Aluminum and steel are two commonly used metals in the construction of electronic equipment, especially in production. In the development stages, copper alloys, such as brass or bronze, are frequently used because of their relative ease of machinability and soldering. Other metals are used in special applications where aluminums, coppers, or steels are not satisfactory.

Metals are available in a variety of shapes and sizes. Standard products and their definitions are:

1. SHEET. A rolled section, rectangular, at least 0.006 and less than 0.250 in. thick.
2. PLATE. A rolled section, rectangular, 0.250 in. thick or thicker.
3. FOIL. A rolled section less than 0.006 in. thick.
4. ROD. A long, solid, round section at least $\frac{3}{8}$ in. in diameter.
5. WIRE. A "rod" less than $\frac{3}{8}$ in. in diameter. Cross section may be any symmetrical shape.
6. BAR. A "rod" which is not round. Cross sections may be square, rectangular, hexagonal, octagonal.
7. TUBE. A hollow "rod" or "bar." Tubing may be drawn or extruded.
8. EXTRUSION. Any other shape that is long with respect to its cross section.

3-1 ALUMINUM

Pure aluminum is soft and ductile and consequently not strong enough for electronic assemblies. However, with small percentages of other metals added, aluminum alloys can be formed which compare favorably with steel for strength and rigidity. The most common additives are copper, manganese, silicon, magnesium, and zinc, although other metals may also be used. Most aluminum alloys weigh about one-tenth of a pound per cubic inch, which is about one-third the weight of steel. In addition, pure aluminum is a relatively good electrical conductor, having a conductivity about six-tenths that of copper. Aluminum alloys have a somewhat lower conductivity because of the lower conductivity of the additives, but in general the conductivity of most aluminum alloys is higher than that of steel or brass. Because of its light weight and good conductivity, aluminum is a preferred metal for chassis and for structural members on all airborne equipment.

Since almost any metal can be added to aluminum in any varying amount, the list of aluminum alloys is virtually endless. A standard nomenclature has been adopted to describe aluminum alloys. Each alloy is designated by a four-digit number which indicates to some extent what additives are contained in the mixture. The first digit of the four-figure number indicates the main additive as expressed in Table 3–1. (All tables will be found in Appendix A at back of book, page 141.) The second digit indicates the degree of control of impurities or alloy modifications. The last two digits in the materials which are designated with a number beginning with 1—the 99% aluminum alloys—are the number of hundredths of a percent above 99% that is aluminum in the material. Thus, 1060 contains 99.60% aluminum, 1100 contains 99.00%, and 1235 contains 99.35%. In the case of the other alloys, beginning with 2 to 8, the last two digits are simply serial numbers and do not indicate anything about the alloy.

When an aluminum alloy is first fabricated, it is not as ductile as pure aluminum, but it still needs further strengthening. Some elements, copper for example, can be hardened and strengthened by heating, and aluminum alloys containing these metals can also be strengthened by heating. These are called *heat-treatable* or *thermally hardened* alloys. Alloys which cannot be strengthened and hardened by heating are *non-heat-treatable*. These, however, can be *work-hardened.*

The temper of work-hardened alloys is designated by a letter following the four-digit number. These suffixes are shown in Table 3–2. Additional digits after the suffix H may be used to indicated the degree of hardening.

The temper of thermally hardened alloys is designated by the suffix T followed by one or more digits. These designations are shown in Table 3–3. Additional digits indicate variations in treatment. For example, suffix T651 designates an alloy which is heat-treated, stress-relieved by stretching, and then aged.

Heat-treatable alloys containing copper or zinc are more subject to corrosion than non-heat-treatable alloys. To improve resistance to corrosion, these alloys may be *clad* with an alloy which is corrosion-resistant. The cladding is less than 5% of the total thickness. Clad alloys are called Alclad. A four-figure number designates the *core material*. For example, Alclad 2024 has a core of 2024 alloy and cladding of 1230. Alloys may be clad on both sides or on one side only.

The properties of an aluminum alloy are determined by the amount and type of additives. The properties affected include electrical and thermal conductivities, corrosion resistance, rigidity, strength, and workability. The physical properties are also affected by hardening the alloy.

Alloys in the 1000 series have no additives, although iron and silicon may be present as impurities (see Table 3–1). These alloys have the best thermal and electrical conductivity, are almost completely resistant to corrosion, and have excellent workability. They have poor mechanical properties, however.

Alloys in the 2000 series have copper as the main additive. When heat treated, these alloys have mechanical properties that compare favorably with steel. However, they are subject to corrosion, so they are usually clad with corrosion-resistant alloys.

Alloys in the 3000 series are non-heat-treatable. These are "compromise" alloys which have moderate strength and moderate workability.

Alloys in the 4000 series have silicon as the main additive, which causes a reduced melting point. Their main application then is as a welding or brazing alloy.

Alloys in the 5000 series have relatively large amounts of magnesium added. These alloys have high strength and good resistance to corrosion. They are easily welded and have reasonably good workability.

Alloys in the 6000 series contain magnesium and silicon and are heat-treatable. They have medium stength, good workability, and good resistance to corrosion.

Alloys in the 7000 series contain zinc and are the strongest of the aluminum alloys. Their exceptional strength is obtained at some sacrifice of workability, corrosion resistance, and electrical properties.

Although there are a large number of alloys in each series, only a few are used extensively. Some popular alloys and their applications in electronic equipment are listed in Table 3–4.

Where weight is a consideration, aluminum is much superior to steel, even in applications requiring stiffness or strength. Figure 3–1 shows the relationship between steel and aluminum for stiffness. Column A shows that for equal thicknesses aluminum is about one-third the weight of steel. In order to attain equal stiffness, aluminum must be thicker than the steel; but as shown

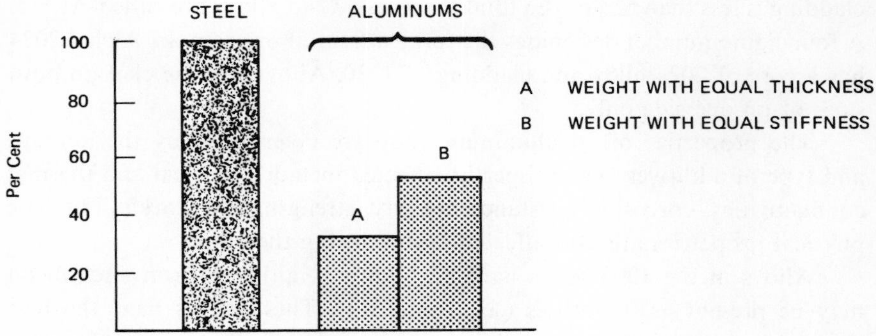

FIGURE 3-1 Weight-Stiffness Comparison

in column B, it will weigh only slightly more than 50% of the weight of steel.

Figure 3–2 shows the relative weight of various aluminum alloys as a percentage of the weight of steel which has equal strength. The 1100 alloy is included to show that even a "soft" alloy weighs less than steel with equal strength. In this case, the aluminum would have to be almost three times as thick as steel of equal strength, thus 1100 is not usually used where strength is required. However, it should be noted that strong aluminum alloys have strength equal to steel with less than half the weight. Aluminum alloys 7075–T6 and Alclad 2024–T3 are actually stronger than steel.

For electrical applications where conductivity is the most important factor, aluminum compares well with copper. Figure 3–3 shows the relative conductivity of selected aluminum alloys as a percentage of that of copper for equal volumes. Although the conductivity of aluminum is lower than

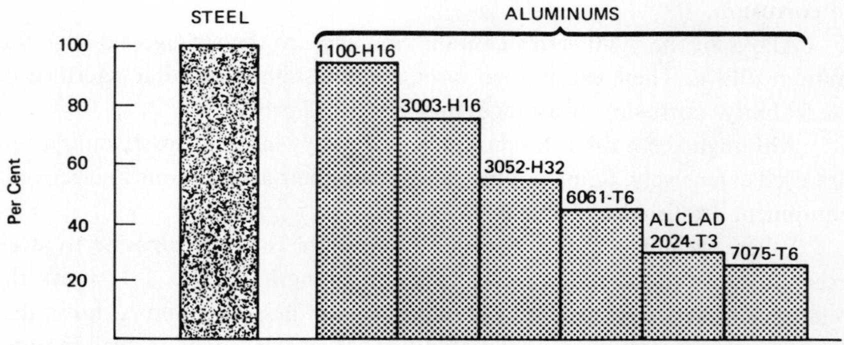

FIGURE 3-2 Weight-Strength Comparison

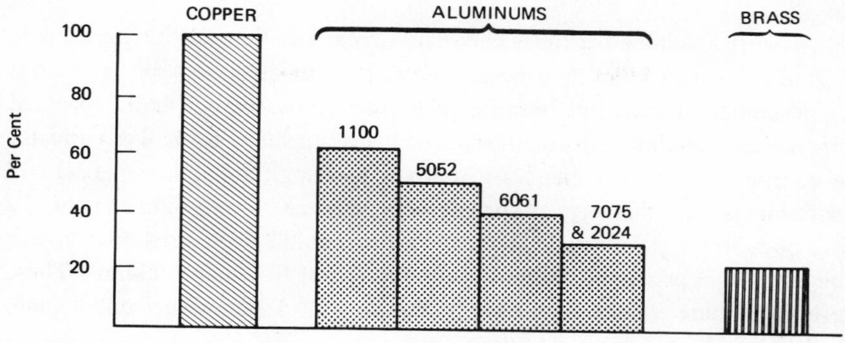

FIGURE 3-3 Relative Conductivity for Equal Volumes

that of copper, it is higher than that of brass, as shown in the figure, and is quite satisfactory in fact. From a weight standpoint aluminum excels. That is, if the cross section of aluminum is increased to a point where the weight of the aluminum piece equals that of the copper piece, the resistance of the aluminum piece may actually be less than that of the copper. This is shown in Fig. 3–4. For equal weights of copper and aluminum 1100, the aluminum has almost double the conductivity of the copper. All of the aluminum alloys shown in Fig. 3–4 can compete quite favorably with copper as far as conductivity is concerned.

The products available in some popular aluminum alloys are shown in

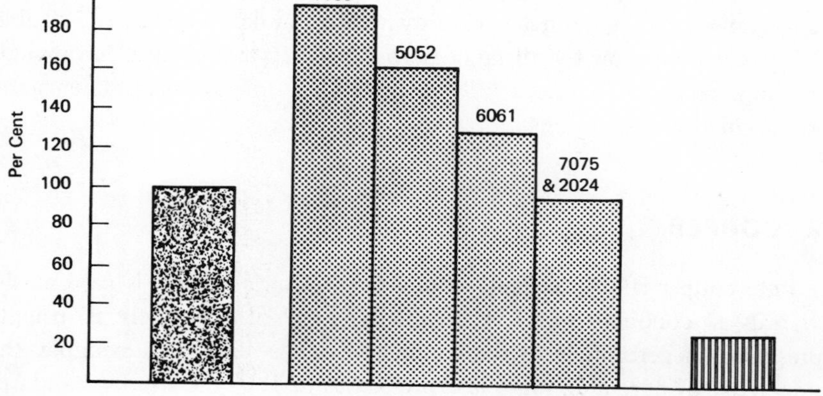

FIGURE 3-4 Relative Conductivity for Equal Weights

Table 3–5. It should be noted that some shapes are not available in individual alloys.

Sheet is available for most alloys in almost any desired thickness up to 0.250 in., at which value it becomes *plate*. Plate is also available in almost any desired thickness, but because of looser tolerances is usually specified at increments of sixteenths or thirty-seconds of an inch. Table 3–6 indicates the commonly used thicknesses available. Although stock may be had in other thicknesses, delivery of special sizes may be slower than that of the more popular sheet and plate stock shown in Table 3–6. Also shown in this table are the approximate weights per square foot for each thickness. These are approximate values, assuming a density of 0.1 pound per cubic inch, which is within 2% for most alloys.

The tolerance specified on the thickness depends on the thickness and width of the product as well as on the stiffness of the alloy. In general, softer alloys are held more closely to the specified thickness than harder ones. Table 3–7 lists tolerances for some common alloys as a function of thickness up to 6 in. and for widths up to 54 in.

Aluminum bar and rod are available in a range of sizes up to 8 inches in diameter for round stock and up to 4 to 10 inches for other shapes. Although round stock in any specified diameter can be obtained, commonly used diameters are multiples of one-eighth of an inch up to one inch in diameter and by quarter-inch increments above this. Square, hexagonal, and octagonal cross sections are more commonly obtainable in multiples of one-quarter inch in distance across flats. Bar stock and rectangular wire are commonly available in an assortment of sizes in which the narrow dimension or thickness is in multiples of one-eighth inch (up to about 3/4 inch) and the wide dimension or width is in increments of a quarter-inch. Table 3–8 lists tolerances on the diameter for drawn wire and cold finished rod; Table 3–9 lists tolerances on the distance across the flats for square, hexagonal, and octagonal cross sections; and Table 3–10 lists tolerances on dimensions of rectangular cross sections.

3-2 COPPER

Pure copper is an excellent conductor of electricity and is used as the standard of conductivity. The conductivity of other metals is usually expressed as a percentage of the conductivity of copper. For example, the conductivity of pure aluminum is expressed as 62% of the copper standard. *Alloys* of copper, however, have much lower conductivities than aluminum.

As with aluminum, copper can be combined with many different metals in varying amounts to form an endless variety of alloys. For convenience, copper and its alloys are divided into nine groups, as follows:

1. COPPERS. These are at least 99.9% pure copper and are excellent electrical conductors. If silver is included as an impurity, it is counted as copper, since the conductivity of silver is about the same as that of copper.
2. BRASSES. Brass is an alloy of copper and zinc. These range from 95% copper/5% zinc to 60% copper/40% zinc. Brass is sometimes designated by the percentage of copper in it. Thus, *80% brass* (or sometimes, brass 80%) contains 80% copper and 20% zinc. A popular alloy is *yellow brass*, which contains 65% copper and 35% zinc.
3. LEADED BRASS. This is a 65% or yellow brass in which a small amount of zinc has been replaced by lead. Leaded brass may contain up to 2.5% lead. The addition of lead makes the material very easy to machine. *Low-leaded brass* is 65% copper, 34.5% zinc, and 0.5% lead.
4. TIN BRASS. A small percentage of tin, usually from 0.5 to 2%, is contained in this alloy. Most common is *naval brass*, which is 60% copper, 39.25% zinc, and 0.75% tin.
5. PHOSPHOR BRONZES. These are alloys of copper and tin and are designated by the percentage of tin in the alloy. The two standard alloys in this group are *phosphor bronze 5%*, which contains 95% copper and 5% tin, and *phosphor bronze 8%*, which contains 92% copper and 8% tin.
6. SILICON BRONZES. These are alloys of copper and silicon and may contain other elements. *High-silicon bronze* contains 3% silicon, and *low-silicon bronze* contains 1.5% silicon.
7. NICKEL SILVER. This is a yellow brass in which part of the zinc is replaced by nickel. It is designated by the percentages of copper and nickel in the alloy. Mixtures run from 65–18, which contains 65% copper, 18% nickel, and 17% zinc; to 65–10, which contains 65% copper, 10% nickel, and 25% zinc.
8. CUPRO-NICKEL. This is an alloy of copper and nickel and is designated by the percentage of nickel in the alloy.
9. BERYLLIUM COPPER. This alloy contains about 2% beryllium, traces of nickel, cobalt, and iron, and the remainder is copper.

Pure copper weighs 0.32 pounds per cubic inch, or about 3.2 times the weight of aluminum. This is slightly heavier than steel. The properties of the copper alloys depend on the elements added, thus the applications of the several alloys differ. Not all are used in electronics.

Since copper is an excellent conductor, it is used in wiring in most electronic systems. Copper wire is available in standard diameters as a solid conductor or as stranded wire. Stranded wire is more flexible than solid wire of the same diameter. Both are used for point-to-point wiring. (Copper-coated rigid cards of insulating material are used in printed wiring boards, which are discussed in Chapter 9.)

Because brass is relatively easy to machine and plate, it is frequently used in engineering models for microwave applications. Its conductivity is only one-fourth that of copper; consequently it is usually silver-plated when used for microwave waveguides and cavities. Its weight varies with the amount of copper from 0.32 pounds per cubic inch for 95% brass down to 0.30 for 60% brass.

Leaded brass is not used in electronic applications. Although it is easy to machine, the lead may contaminate plating solutions. Without plating, leaded brass is a poor conductor at microwave frequencies.

Tin brass is stronger than other brasses and has better resistance to corrosion, but is also a poor conductor. It is rarely used in electronic equipment. Both tin and leaded brasses weigh about 0.305 pounds per cubic inch.

Phosphor bronze is used for springs, metal diaphragms, and contacts in electrical applications. This alloy has the following advantages over steel for these applications: non-magnetic, greater resistance to corrosion, more easily formed. Phosphor bronze weighs about 0.32 pounds per cubic inch.

Silicon bronzes have good corrosion resistance and high strength and are easily welded. They are used in the manufacture of storage tanks. Nickel silver is used for parts made by drawing, spinning, or stamping. Both silicon bronze and nickel silver are *not* used in the electronics industry.

Cupro-Nickel is a strong, ductile alloy with excellent resistance to corrosion. It is used for baffles and heat exchangers in electronic systems. Cupro-Nickel weighs 0.32 pounds per cubic inch.

Beryllium copper is an old alloy which was formerly used for spring applications; however, it is difficult to form and machine and must be aged after every machining or heating operation. It has largely been replaced by phosphor bronze.

In summary, then, copper is used for wiring. Brass has some application for microwave devices. Phosphor bronze is used for springs and electrical contacts. Cupro-Nickel is used for heat exchangers. The other copper alloys are not generally used in electrical or electronic systems. No copper alloys are used for structural members, since steel is stronger and aluminum is lighter. For the designer, copper's most important application is for wiring. This is covered in detail later, in Chapter 8 on hard wiring, and in Chapter 9 on printed wiring.

Brass is available in all configurations: sheet, plate, wire, rod, bar, tubing, and extrusions. Hollow rectangular extrusions are used as wave guides at microwave frequency. Standard waveguide sizes are covered in Chapter 12 on microwaves. Standard brass rods are available from 0.049 in. to 6 in. in diameter. Brass bars are available in square, hexagonal, and rectangular shapes in sizes up to 3 in. Popular sizes have dimensions which are multiples of a quarter-inch. Table 3–11 lists commonly available sheet sizes and tolerances on thickness.

3-3 STEEL

Steel is an alloy of carbon and iron and usually contains additional metals, notably manganese. Steel is relatively inexpensive and much stronger than iron and is therefore a preferred material for structural applications.

Carbon steel has properties controlled by its carbon content. In general, the greater the amount of carbon, the stronger is the steel. This carbon content may range from less than 0.1% to as high as 5%. The metal is designated as *low-carbon steel, medium-carbon steel,* or *high-carbon steel,* according to the carbon content. For electronic equipment, *low-carbon steel* is sometimes used for chassis and brackets; higher carbon steels are rarely used since they are more difficult to fabricate. It should be noted that carbon steel contains other elements (frequently in higher percentages than the carbon in the steel), but the material is called carbon steel if its characteristics depend on the carbon rather than on the other metals in the alloy.

Alloy steel has distinctive properties or characteristics which are caused by some added metal other than carbon, or by another metal and carbon together. Alloy steel is usually referred to by the name of the important alloy, as *nickel steel, chromium steel,* etc. Frequently an alloy steel may contain more carbon than the designated metal, although its special characteristics derive from the other metal. The amount of the designated metal in alloy steel may vary from a trace to about 5%. Most alloy steels are medium-carbon steels which are modified by the additive.

The principle disadvantage of carbon steel is its vulnerability to corrosion. Addition of chromium or nickel inhibits corrosion, and thus chromium steel and chromium-nickel steel are called *stainless steels.* Stainless steels are designated by a three-digit number, in which the first digit indicates the type and the other two digits are merely serial numbers.

Chromium steel is designated by a 4 for the first digit. *Hardenable* chromium steels contain from 12 to 18% of chromium and about 0.1 to 1% of carbon. The basic type is 410, which contains 12% chromium and less than 0.15% carbon. These stainless steels may be hardened by heat treatment. By increasing the carbon content, the strength can be increased. Thus, type 420 contains more than 0.15% carbon and is stronger than type 410.

By increasing the chromium content to the range of 14 to 27% and lowering the carbon content, corrosion resistance is improved, but the resultant chromium steel is *nonhardenable* by heat treatment. The basic type is 430, which contains 17% chromium and about 0.10% carbon. All stainless steels in the 400 series, both hardenable and non-hardenable, are magnetic. They have good machining properties and can be cold-formed.

Chromium-nickel steel is designated by a 3 for the first digit. These stainless steels have excellent corrosion resistance. The basic type is 302, which contains 18% chromium and 8% nickel. The chromium-nickel steels are not heat treatable but may be hardened by cold working. Hardening by cold working can be increased by lowering the chromium and nickel content. Reducing the carbon content improves corrosion resistance. Type 301 contains about 17% chromium and 6% nickel and by hardening is stronger than type 302. Type 304 is a popular stainless steel. Its improved corrosion resistance is due to lower carbon content: less than 0.08% as against 0.15% in type 302.

During a period when nickel was scarce, some of the nickel in type 302 was replaced by additional manganese to form *chromium-nickel-manganese steel*, which is designated by a 2 for the first digit. Since all stainless steels contain manganese (up to 2% in the 300 series and up to 1% in the 400 series), these in the 200 series are sometimes erroneously called simply chromium-nickel steels. The basic type is 202, which contains 18% chromium, 5% nickel, and about 8.5% manganese. Chromium-nickel-manganese steels are not heat treatable but, like chromium-nickel steels, may be hardened by cold working. Work hardening can be improved by lowering the content of the three principal alloys. A popular type is 201, which contains 17% chromium, 4.5% nickel, and about 6.5% manganese. Stainless steels in the 200 series have excellent corrosion resistance. Those in the 200 and 300 series that contain nickel are nonmagnetic, but may become slightly magnetic when worked.

The weight of carbon steels is approximately 0.28 pounds per cubic inch. Stainless steels in the 400 series, the chromium steels, weigh about the same. Chromium-nickel steels in the 200 and 300 series are slightly heavier, weighing about 0.29 pounds per cubic inch.

Sizes of steel sheets and plates are usually designated by a *gauge number* rather than a thickness. Gauge numbers are used to designate wire sizes, hoop sizes, and other metal parts, and different lists are used for different materials. The applicable list of gauge numbers for steel sheets and plates is called United States Standard Gauge (abbreviated, U. S. Std.). This is officially a weight gauge. The gauge numbers and corresponding weights of steel sheets or plates are shown in Table 3–12. Also included is the thickness of steel which will have approximately the weight indicated. The thicknesses listed correspond to the weights for carbon steel. The same gauge numbers apply to stainless steel, but it is common to order and specify stainless steel by thickness rather than by its gauge number.

In electronic equipment, cold-rolled low-carbon steel may be used for cabinets and structural supports where corrosion is not a problem. Commonly used sheet sizes run from gauge number 11 to gauge number 30. Where corrosion is a problem, stainless steel may be used in these applications and also for chassis. Aluminum is used where weight is a consideration, as in airborne applications, but steel—even stainless steel—is much less costly.

Stainless steel and carbon steel are also available in all the standard products listed on page 17. Special extrusions and castings may be used for structural supports in electronic equipment where close tolerances are not required.

3-4 SPECIAL METALS

For certain applications, special metals are required. These may be alloys which have been designed for some desired characteristics or elements which are unique.

Special iron alloys have been designed for magnetic shielding applications. A common material for this purpose is *mu metal.* Other similar alloys developed more recently are *netic* and *co-netic*, both of which are somewhat superior to mu metal. Since investigation in magnetic shielding is a continuing development, the designer should consult manufacturers' catalogs for the latest specifications of such materials.

Magnesium is a very light metal, weighing less than two-thirds the weight of aluminum. In applications where light weight is of major importance the designer should consider using this material but should be aware of its disadvantages. Special precautions are necessary during machining and fabrication since magnesium chips and dust are highly inflammable. Also, this metal requires special preparation for joining. Alloys of magnesium with small amounts of aluminum, zinc, and manganese are used for sand castings and extrusions.

3-5 CHOICE OF MATERIAL

The selection of a metal or metals for a production design is based on many factors. The object is to achieve the desired quality with minimum total cost. Workability, weldability, and weight may be more important factors than initial cost, since the cost of manpower is usually greater than the cost of material.

During engineering and preproduction stages, frequent changes and adjustments are made, and it is necessary that they be made quickly and easily by a technician. Here the selected material might be brass, since it is easily machined and can be soldered with an iron or a light torch. Although brass will not be the material used in production, its use during these early stages saves time for engineers and technicians.

In production, a savings in time may not seem so dramatic, but it can be more appreciable in fact. A saving of only one minute per unit can amount to several man-hours, and consequently several dollars saved in a large run.

For chassis, brackets, and structural parts, aluminum is preferred because of its light weight, good conductivity, and relatively good strength. Carbon steel and stainless steel are also used for good strength at low cost where conductivity and weight are not important. Occasionally, magnesium is the choice where weight is the most important consideration. The physical and mechanical properties of these materials are the important factors affecting the selection. These properties are shown in Table 3–13 for some common aluminum alloys and magnesium, and in Table 3–14 for some common stainless steels and also low-carbon steel.

Corrosion resistance, workability, and joinability are all important considerations in selection of metals. Joining includes soldering, brazing, and welding. Workability includes all fabrication processes such as forming, bending, spinning, and machining. Corrosion resistance will be covered in detail in Chapter 7, but relative corrosion resistance of different metals is included here. Tables 3–15, 3–16, and 3–17 show the fabricational properties, joinability, and general corrosion resistance of aluminum alloys, copper alloys, and steels, respectively. In these tables the qualities are rated from A to D, A being excellent, B good, C fair, and D poor or unacceptable.

Having decided on a metal, the designer must then select a size and shape. If possible, he should choose a size which has been used before or is being used on another job and which is easily obtainable. This uses up what is left over from other jobs and also simplifies stockpiling. The designer should try to minimize the number of different materials and different sizes which are used on various jobs. This was covered in Sec 2-2 on preferred stocks.

The selected sizes and materials become *standard stock* or *preferred sizes* when they are used frequently, and future selections should be made from standard stock, if possible.

It sometimes happens that the size or shape needed can be cut from two or more different original products. For instance, a piece of aluminum $\frac{1}{8}$ in. $\times$ 2 in. $\times$ 12 in. can be cut from sheet or bar stock. If both are on the standard stock list, the designer should select whichever will require the least machining or handling.

4

SHEETMETAL DESIGN

Sheetmetal work is required extensively in cabinets, chassis, boxes, and some structural members of electronic systems. In designing these parts for production, cost is an important consideration. Since labor costs are more expensive generally than material costs, the design should incorporate methods and techniques which reduce handling. It is possible to go to extremes to avoid extra operations, such as specifying solid silver as a material in order to avoid a silver-plating operation, but this would hardly result in a cost reduction. However, a few extra pennies for materials or for set-ups on numerically controlled machines can sometimes save dollars in labor or overhead. The potential savings from using cheaper material or techniques must always be weighed against labor costs.

Sheetmetal operations do not include assembly and wiring, but the shape of a finished cabinet or chassis will nevertheless affect these later operations. The designer must be aware of the whole equipment, since it is possible that the most economical chassis will lead to more expensive assembly and wiring requirements.

4-1 CLARITY

To avoid costly rejections of fabricated products, the designer must assume that machinists may be *careless*. It is not enough to have a design which is technically correct; the design must be foolproof. The flange drawing shown partially in Fig. 4–1 is from an actual case which illustrates this point. The flange is square, but the positions of the holes are *not* located on the corner of a square. Although the dimensions are clearly marked, the parts would sometimes come back with the holes incorrectly positioned exactly on the corners of a square. If this happened once, it would be only a machin-

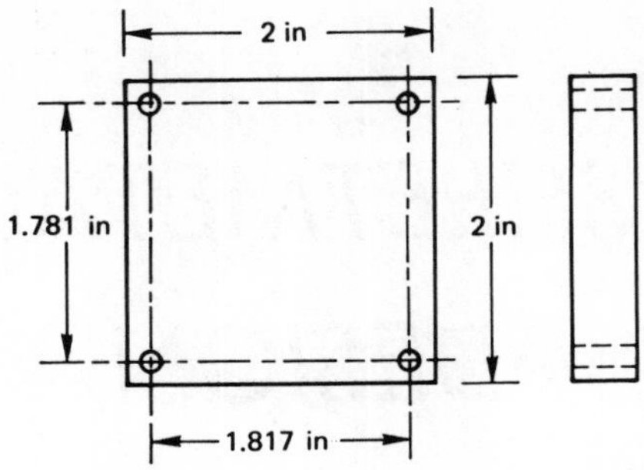

FIGURE 4-1 Flange

ist's error, but the fact that it happened time and again indicates that it is a poor design. The designer was not considering the *careless* or tired machinist who sees the four holes apparently on the corner of a square and looks at only one dimension. The fact that the two dimensions, 1.781 and 1.817, had the same digits was not an underlying cause of the error. These were changed to 1.782 and 1.816 with appropriate changes in tolerances, but again the error occurred. Errors were eliminated entirely when a note pointing to the

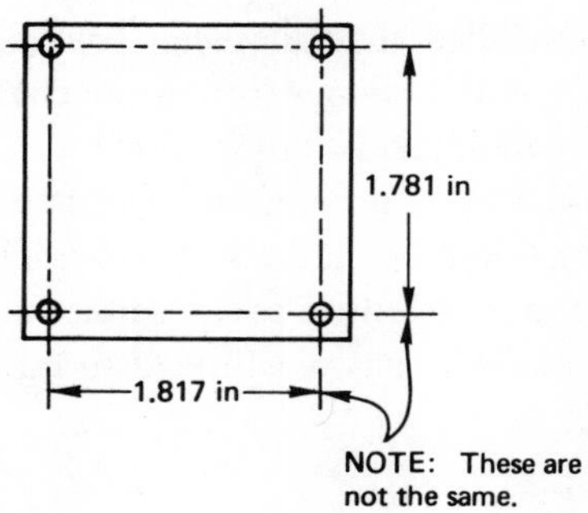

FIGURE 4-2 Foolproof Design

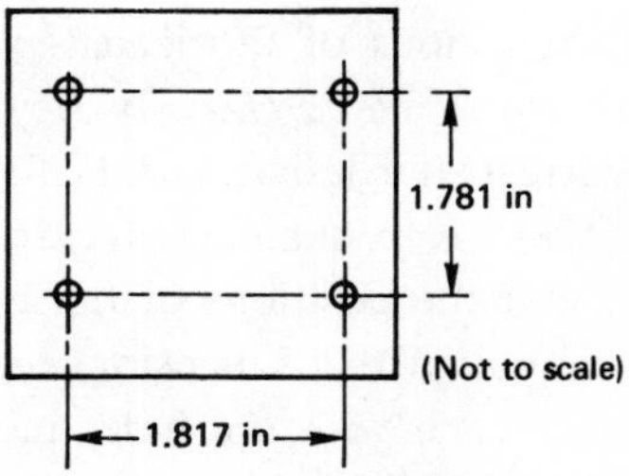

FIGURE 4-3 Design Solution

discrepancy was added as shown in Fig. 4–2. Technically, the added note is redundant and therefore wrong. Actually it eliminated a source of errors. A more esthetic solution, adopted later, is shown in Fig. 4–3. The holes in this figure are obviously not on the corners of a square, and the machinist is thus alerted to look for two different dimensions.

4-2 BENDING

Designing bends in sheetmetal is not an exact science, but the large body of experience has furnished several rules of thumb. The parts of a bend are shown and defined in Fig. 4–4. The *radius* of the bend is the inside radius of curvature. The *angle* of bend is the angle through which the flange was bent. The most common bends are 90°, but greater bends are possible, as shown in this illustration. In fact in some applications 180° bends are used. The shorter of the two arms of a bend is called the *flange*, and the longer is the *web*. The *length* of the bend is simply the width of the bent piece. In the figure, the length is less than the flange and the web, but a bend can be made in a wider sheet also. The thickness of the metal is not part of the bend, but it is an important factor in the design.

It is possible to make bends with any angle between 0° and 180°, but 90° bends are preferred. The angle of bend should always be indicated on a drawing, even for 90°, but if it is omitted the bend is assumed to be 90°. Some radius should always be indicated because sharp bends with no inside radius

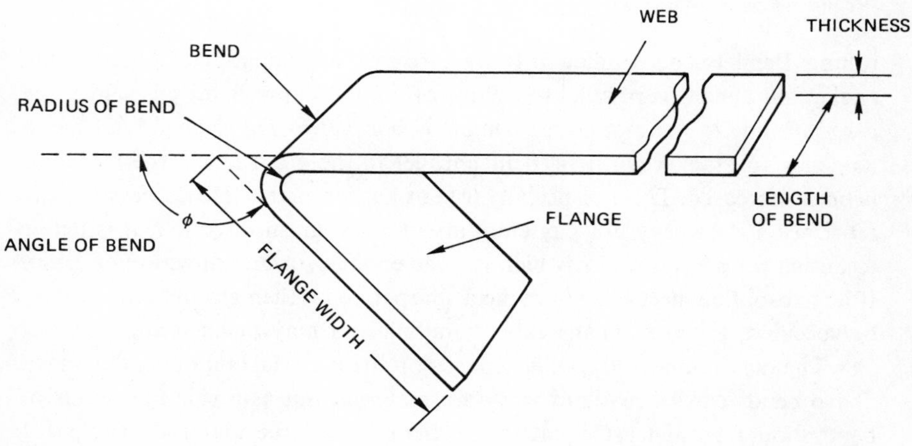

FIGURE 4-4 Parts of a Bend

are difficult to make and are too fragile. The radius should be as large as possible. In general, harder materials require a larger bend radius than softer ones. The grain direction also affects the size of the minimum radius of bend. In Fig. 4–5, three different bends are shown in relation to the direction of

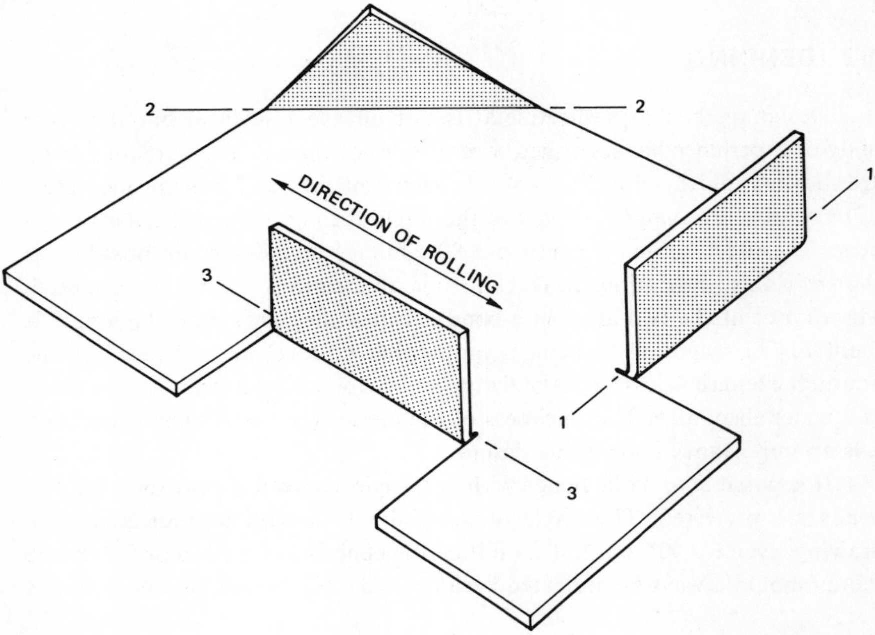

FIGURE 4-5 Axes of Bends

rolling. Bend 1 is perpendicular to the direction of rolling, bend 2 at an acute angle, and bend 3 is parallel to the direction of rolling. Bend 1 is said to be *perpendicular to the grain*, and bend 3 is *parallel to the grain*. Metal has a tendency to crack when it is bent parallel to the grain unless the radius of bend is increased. This is especially true of harder metals. However, since the fabrication shop may not pay close attention to grain direction, it is better to design for a bend radius which is large enough for any direction of grain. If it is absolutely necessary to make a sharper bend, then grain direction may be specified, but this means extra handling and may mean more waste.

The maximum bend radius depends on the material and on its thickness. Sharp bends can be made in very thin material, but as a rule the radius of bend should be at least equal to the thickness of the material. Table 4–1 lists the *minimum* bend radii for several thicknesses of some popular metals. In each case the bend radius is given as a multiple of the thickness of the material. The values in Table 4–1 are for bends parallel to the grain and may be used for bends in other directions also.

The width of the flange should be at least $\frac{1}{4}$ inch for thin material. For softer alloys like 1100–0 aluminum or 5052–H32 aluminum, the minimum flange width should be about four times the thickness of the material. For harder materials like aluminum alloy 6061–T6 or the 400 series of stainless steels, this should be increased to about eight times the thickness of the metal.

The *bend line* may be thought of as the line where the bend starts. It is the line at which the web or the flange joins the curved bend. This is shown in cross section in Fig. 4–6. If a hole in the metal is too close to the bend

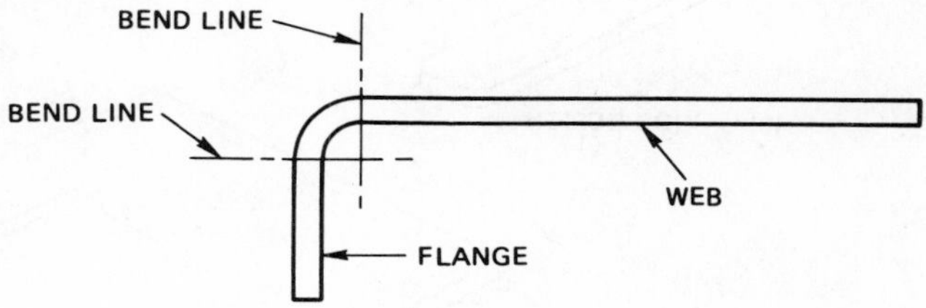

FIGURE 4-6 Bend Line

line, it may be distorted during bending. To prevent this, the distance between the hole and the bend line should be at least 1.5 times the thickness of the material. If it is absolutely necessary to locate a hole closer than this to the bend line, the hole must be drilled or punched after the metal is bent. This is more difficult and more expensive than punching the hole in unbent flat sheet and should be avoided.

When two or more bends are to be made in the same piece of material, the minimum distance between adjacent bends should be ten times the thickness of the metal. This is shown in Fig. 4–7.

Relief notches may be used near bends to prevent fracturing the surface

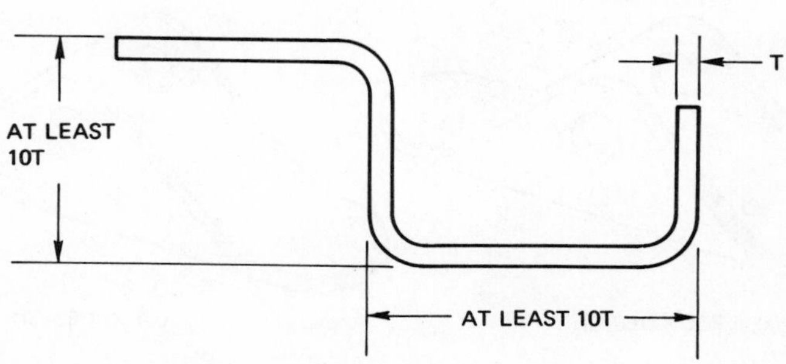

FIGURE 4-7 Multiple Bends

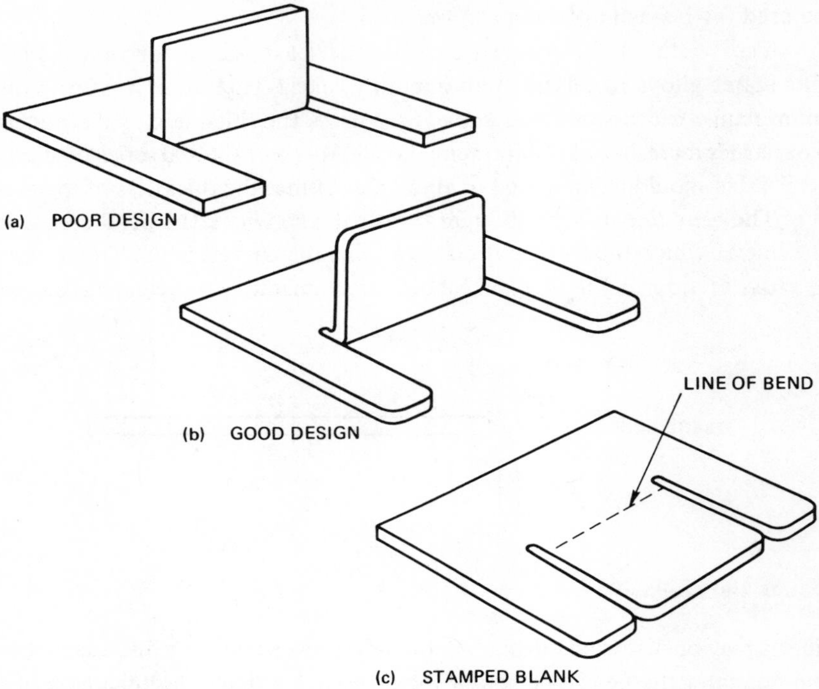

FIGURE 4-8 Relief notches to prevent cracking

close to the bend or to permit bending in otherwise impossible places. Examples are shown in Figs. 4–8 and 4–9. In Fig. 4–8(a), the bend cannot easily be made to close tolerances without distorting the flat portions of the part. In Fig. 4–8(b), the relief notches permit the bend to be made easily. These notches should extend at least twice the stock thickness beyond the

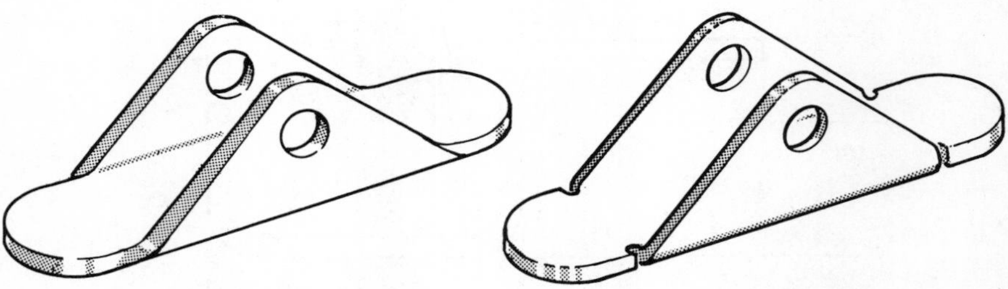

FIGURE 4-9 Relief Notches to Permit Bending

bend. If the part is a stamping, the notches may be stamped at the same time, as shown in (c). Rounded corners are shown since they permit a simpler die construction. Fig. 4–9(a) shows a part which is difficult and very expensive to fabricate. Using relief notches, as shown in (b), the part can be made more easily and economically.

4-3 STRENGTHENING

Additional structural supporting members may be used to strengthen sheetmetal surfaces but add to the cost and weight of the final product. In many cases satisfactory strengthening can be achieved by stamping beads and flanges into the metal. Simple beads are shown in Fig. 4–10. These are

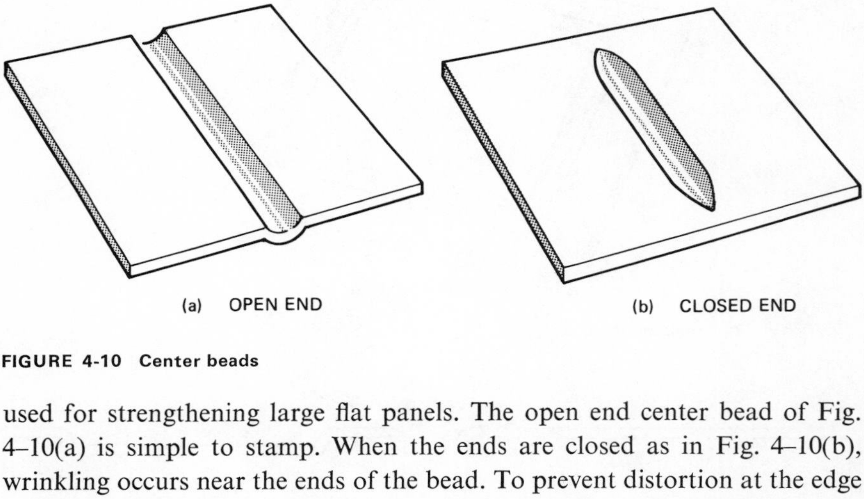

(a) OPEN END (b) CLOSED END

FIGURE 4-10 Center beads

used for strengthening large flat panels. The open end center bead of Fig. 4–10(a) is simple to stamp. When the ends are closed as in Fig. 4–10(b), wrinkling occurs near the ends of the bead. To prevent distortion at the edge of the sheet from this wrinkling, the end of the bead must be no closer to the edge of the metal than 25 times the thickness of the metal. Where greater rigidity is desired, intersecting beads may be used as shown in Fig. 4–11. The die for this type of bead is more expensive than that for Fig. 4–10.

Instead of a straight bead, a continuous bead with rounded corners may be used as shown in Fig. 4–12. The radius of each corner must be large, preferably equal to the width of the bead. Sharp corners here cause weakness in the metal. Almost as strong—and more economical to produce—is the design of the continuous bead without corners. This design consists of four closed-end straight beads arranged in a rectangle with their ends separated by about the width of a bead.

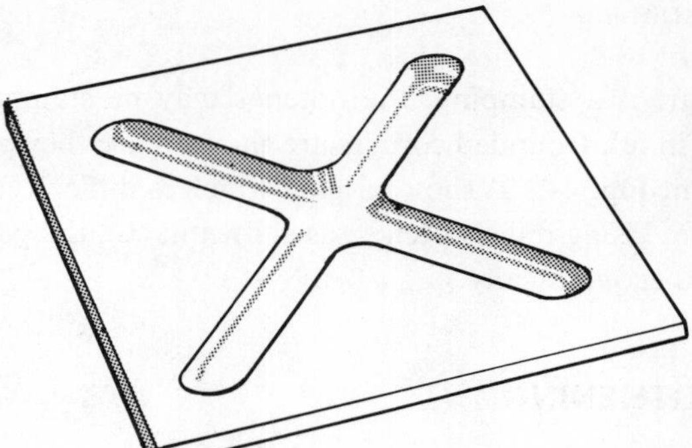

FIGURE 4-11 Intersecting beads

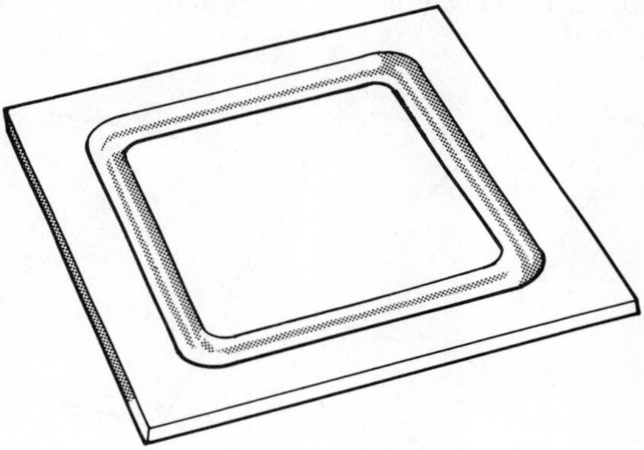

FIGURE 4-12 Continuous bead

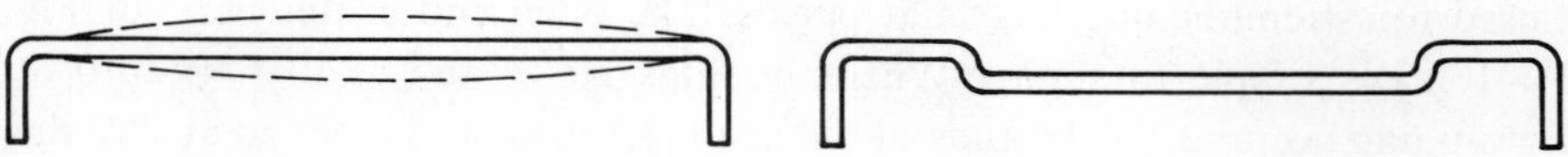

(a) OIL CANNING (b) PREFERRED DESIGN

FIGURE 4-13 Flange

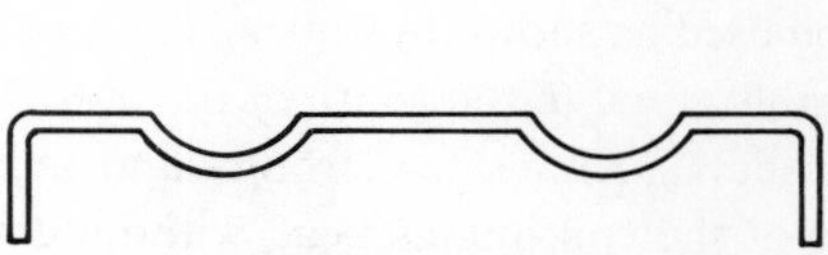

FIGURE 4-14 Flange and beads

A flange may be used to strengthen a flat surface as illustrated in Fig. 4–13(a). However, "oil-canning" can occur as is shown. Depressing the surface as shown in (b) eliminates oil-canning. A flange in combination with beads provides the greatest rigidity, and when both are used, they should project from the same panel surface, as shown in Fig. 4–14. This permits a more economical die design, and both flange and beads can be stamped in one operation.

4-4 PUNCHING

Many of the sheetmetal surfaces in a piece of electronic equipment have a variety of holes and openings. Some of these are mounting holes for meters, connectors, and potentiometers; some may be for ventilation or for weight reduction, and some may be passageways for wiring. Round holes can be drilled simply but holes of other shapes are difficult to make unless they are *punched* and this requires a die for each shape of hole or group of holes. Special dies are expensive; however, standard dies exist to punch odd–shaped holes for mounting standard connectors, meters, and other components.

If a metal sheet or plate is put on a punch to have odd-shaped holes punched into it, it is convenient to have all the round holes punched also instead of drilled. (Even when all the holes are round, it is usually faster to punch the holes than to drill them.) However, the punch diameter should not be smaller than the thickness of the metal. The largest punch size which can be used depends on the thickness and hardness of the material and on the tonnage of the punch press. The maximum length of cutting edge (perimeter) is given by the equation:

$$L = \frac{2000P}{TS} \tag{4-1}$$

where L is the length of cutting edge in inches,

P is the capacity of the press in tons,

T is the thickness of the metal in inches,

and S is the shear strength of the metal in pounds per square inch.

Numerically-controlled punch presses are available which can hold a variety of punches. The metal to be punched is moved automatically to the proper position, and the proper die is moved automatically also. A typical numerically-controlled press may have 24 stations and a pressure of 25 tons. This means that 24 different punches can be used without changing the set-

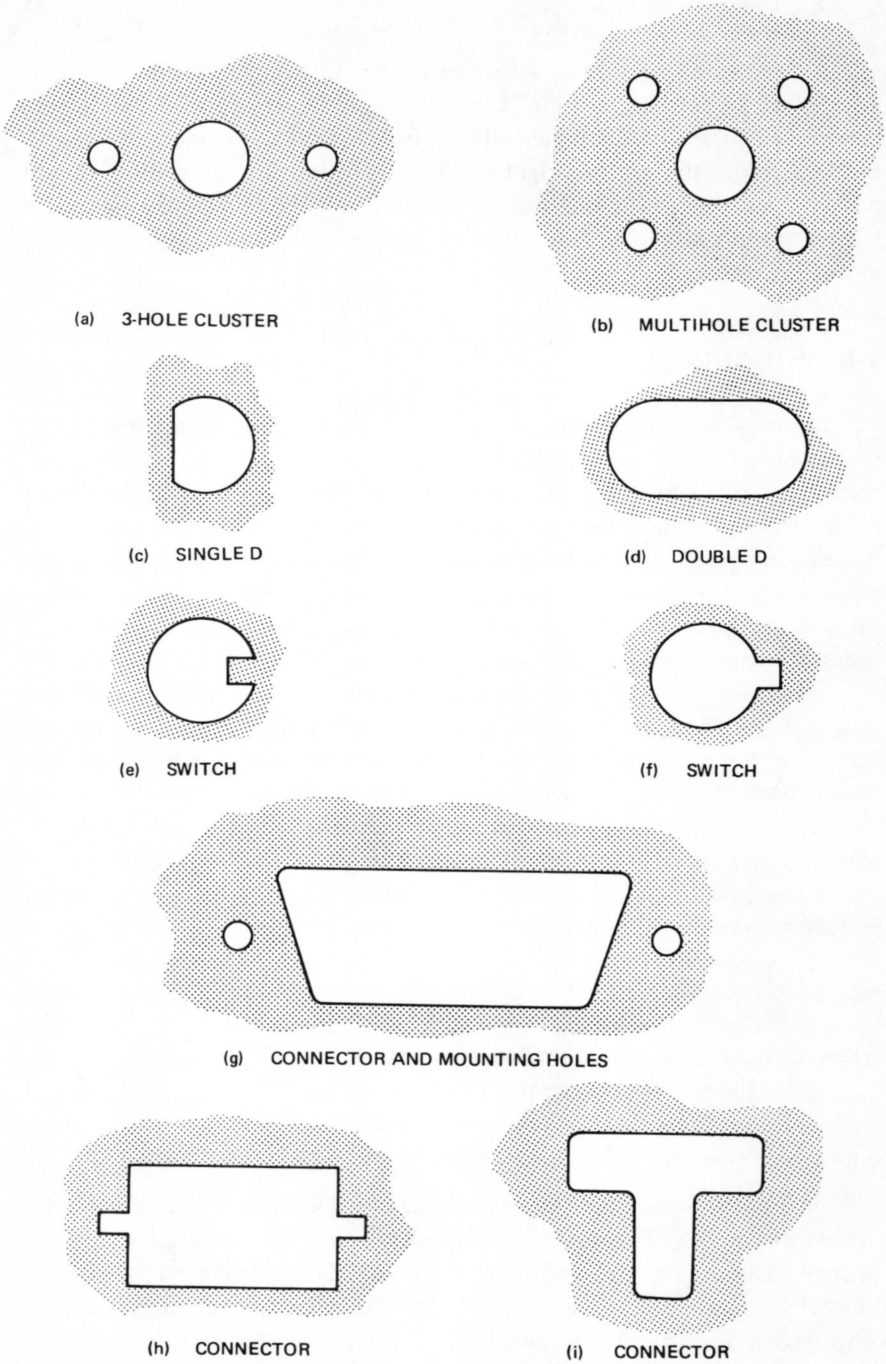

FIGURE 4-15 Standard punches

up. In designing sheetmetal parts, the designer should try to utilize existing dies. If the *company standard list* is limited so that the number of dies needed for connectors and other components does not exceed the stations on the press, the dies can remain in the press permanently and set-up time can be decreased.

Round dies are available for almost any diameter hole from less than $\frac{1}{16}$ in. up to several inches. Rectangular and triangular dies are also available in a variety of standard sizes. Figure 4–15 illustrates some additional odd-shaped dies that were designed for special purposes and which have now become standard because of frequent use. Hole clusters, illustrated in Fig. 4–15(a) and (b) may be had with all holes the same size, or with many different sizes and with from two to 24 holes. The single D and double D punches shown in (c) and (d) are available in a variety of dimensions, as are the holes for mounting switches, shown in (e) and (f). The remaining punches shown (g, h, and i) were designed for mounting holes for specific connectors. As new connectors appeared on the market, special dies were made to simplify cutting the mounting holes. Now most new connectors are shaped to fit into "standard shaped" holes, that is, holes for which dies already exist. There are many other standard dies on the market in a variety of shapes and sizes, so it is difficult today for a designer to justify the use of a special die. More importantly, a designer should strive to limit his need for dies to those in the company's own inventory.

4-5 CHASSIS

The chassis serves as the mounting base for electrical and electronic components. An electronic system may contain only a single chassis, as in a simple automobile radio, or many chassis of different shapes and sizes, as in complex radar or communication equipment. The shape, size and complexity of individual chassis are limited only by a designer's ingenuity. Some typical shapes which have been used in the past are shown in Fig. 4–16. A designer is not restricted to these styles and indeed may choose any shape which will best satisfy the requirements of location and juxtaposition of individual components. However, the chassis shape selected must be one which can be fabricated with standard, available tools.

In choosing a chassis style, the designer must bear in mind the number and kinds of components which will be mounted on it. The general principles mentioned in Chapter 1, Sec. 1.3, apply to chassis design particularly. Human factors must be considered. There must be room for an assembler's hands

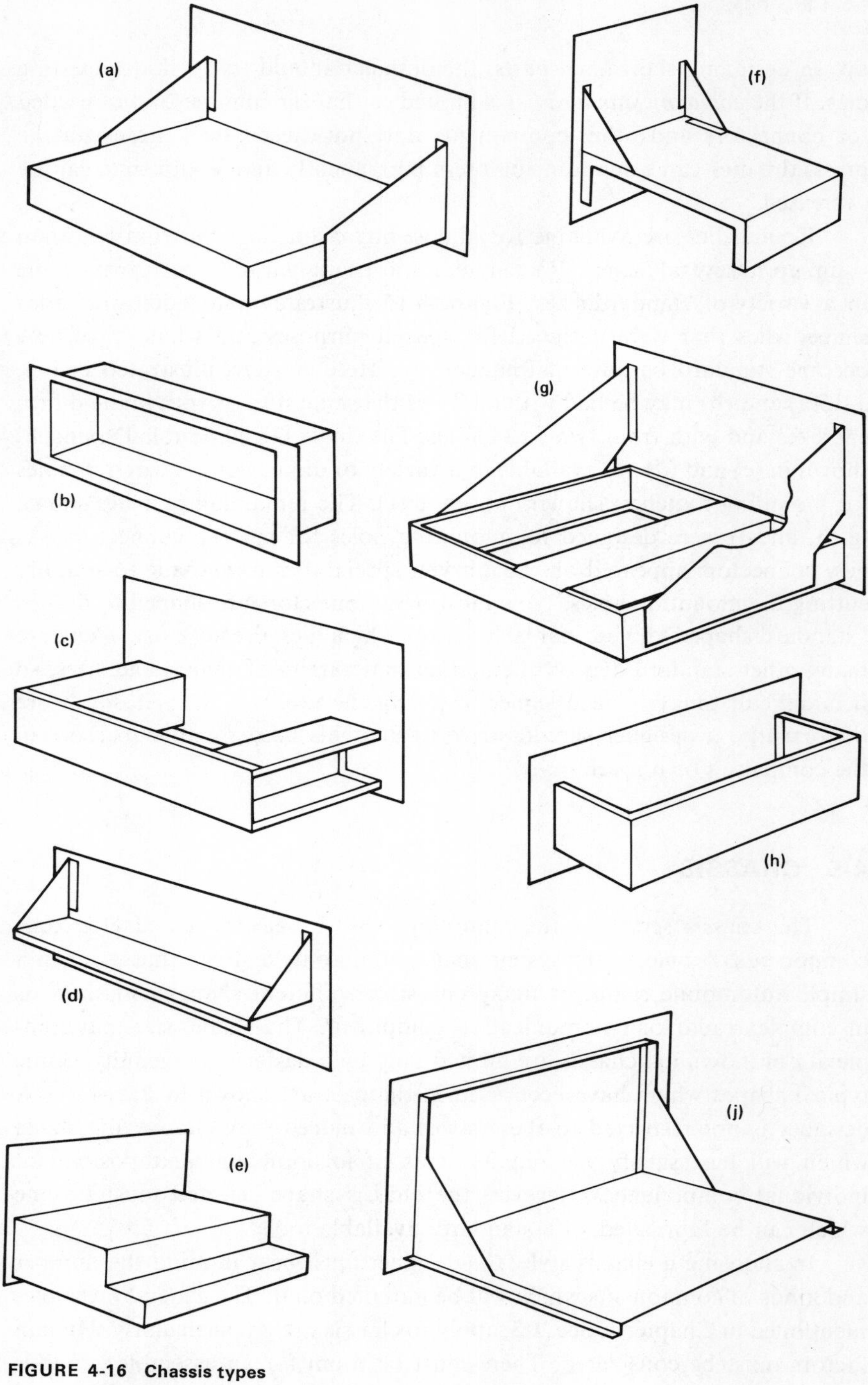

FIGURE 4-16 Chassis types

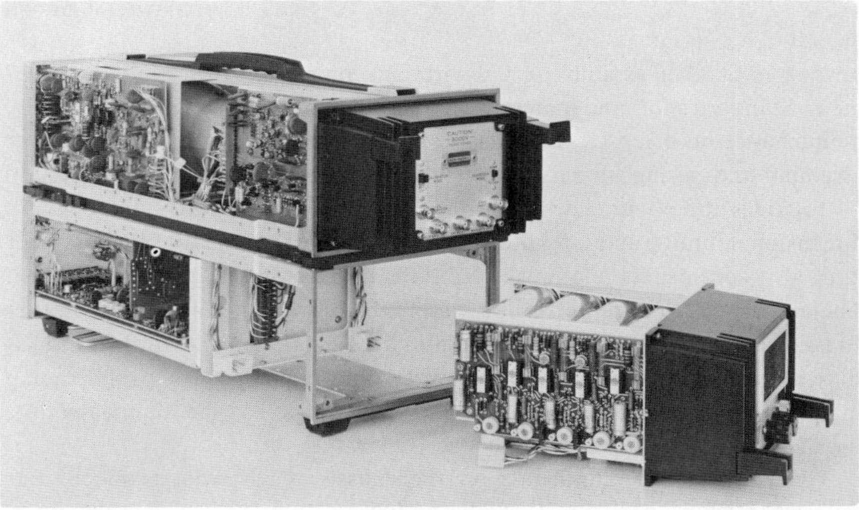

FIGURE 4-17 Plug-in power supply (Courtesy of Hewlett-Packard)

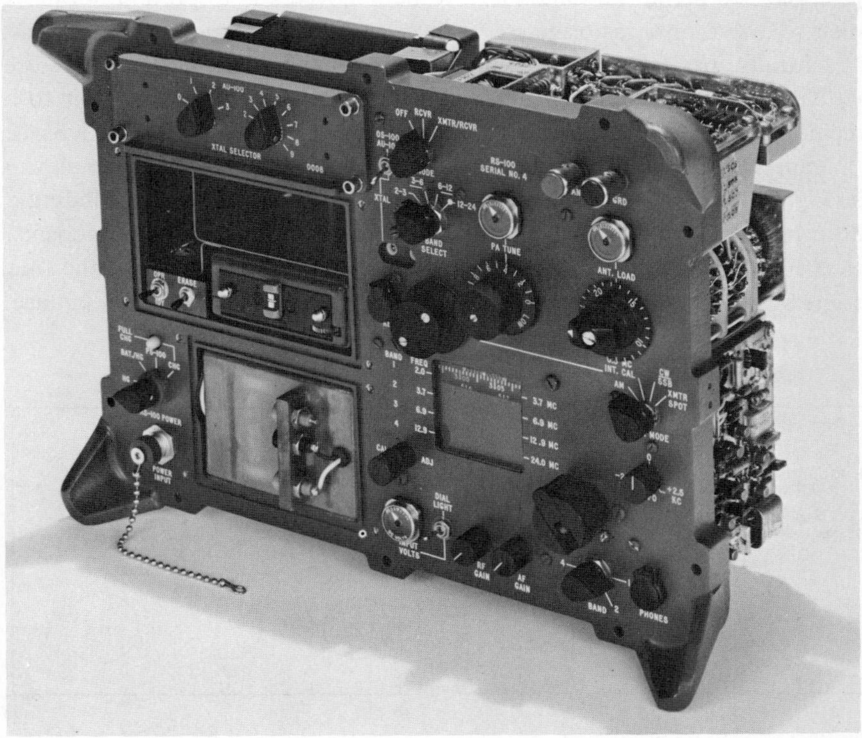

FIGURE 4-18 Cast front panel (Courtesy of GTE Sylvania Incorporated)

or tools to fit when assembling and wiring, or repairing and testing the equipment. Standard tools and materials should be used in construction, assembly, maintenance, and repair. The chassis must be so designed that subsequent assembly and wiring operations can be performed easily and economically.

Besides the obvious design problems of selecting size and shape within dimensional limits and of specifying material for a chassis, the designer must also consider its relationship to different chassis within the equipment. The chassis may be mounted in a cabinet or relay rack alone or with other chassis, or it may be a plug-in module in a larger framework. In Fig. 4–17 is shown a rear view of an oscilloscope with plug-in power supply. By placing the power supply on a separate chassis which plugs into the main assembly, servicing and assembly are simplified.

The layout of components on the chassis is a special problem. For electrical reasons, parts must usually be located to keep lead lengths short. Mechanical considerations dictate positions of heavy components such as transformers, however. Thermal problems must also be considered and parts positioned to permit unwanted heat to be removed and to prevent hot parts such as tubes or transformers from affecting the operation of the circuit.

Ideally, parts should be mounted so their labels or part numbers are visible and legible without disturbing other parts and without danger to a repair man. Test points must be available without disturbing fixed parts.

Special problems require special solutions. The military receiver shown in Fig. 4–18 had to meet stringent shock and vibrations tests, including a drop test. The front panel is a casting which supports all the components, so that in effect the front panel is the chassis. Protruding lugs at the four corners protect knobs and controls if the instrument is dropped on its face.

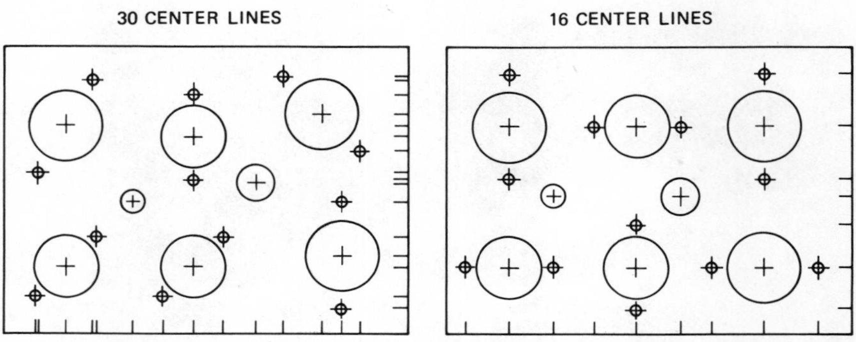

FIGURE 4-19 Hole Layouts

Holes and cutouts in the chassis are necessary for mounting components, and they affect the cost of fabrication. To minimize production costs, all holes should be punched with standard dies, and the number of different sizes kept to a minimum. If possible, holes should be symmetrically placed with as few center lines as possible. Even with automatic, numerically-controlled punch presses, fewer hole sizes and center lines reduce the costs of set-up and programming.

Figure 4–19 illustrates the difference between a poor arrangement of holes and an acceptable one. In Fig. 4–19(a), thirty center lines are required, whereas only sixteen are needed in the preferred arrangement (b).

4-6 PANELS

The panel or front of the electronic system is the interface between the user and the equipment. Here more than in any other area of design, *human factors* are important. The front panel contains the switches and controls which the user manipulates to operate the equipment, as well as the meters, which indicate how the equipment is performing. In some systems, such as those designed for military applications, efficiency of operation is of prime importance, and esthetic considerations, such as style, are nonexistent. On the other hand, stereos, TV sets, and other equipment designed for the mass consumer market must have eye appeal as well as operational simplicity.

The general rules mentioned for chassis design apply also to panels. The number of different holes should be held to a minimum, and common center lines should be used. In addition, the panel should have visual appeal. It should look clean and uncluttered, and markings should be clear and legible. Controls and meters should be grouped in visual balance, although symmetry is not necessarily desirable. If only a few controls are needed on a large panel, they should be located in a group in the center or near an end. It may seem impossible, but when controls are widely separated, a user under pressure (such as in an emergency) sometimes can't locate the proper switch or knob.

Human factors must be carefully considered. Each control should be identified by a legible marking on the panel, and there must be adequate spacing between adjacent knobs so the user will not accidentally move one while his hand is on another. Likewise, if carrying handles are supplied, they should be far enough away from controls to prevent accidental movement.

Where several controls are to be used sequentially, they should be mounted adjacent to one another in the order of operation. Meters should be easily read and should be close to the control which adjusts the voltage or current appearing on the meter. If several meters are used, they should be arranged so that the user's eye moves naturally from one to the next, so that reading the wrong meter is prevented.

A well-planned front panel is shown in Fig. 4–20. The equipment illustrated is a cassette tape player for the consumer market and must be operable by non-technical users. The combination of push-buttons, toggle switches, and slide controls is pleasing to the eye as well as easy to use. A plus is the fact that slide controls which are frequently adjusted in pairs are arranged so that each pair may be moved together with one hand or separately.

FIGURE 4-20 Good Panel Design (Courtesy of Ampex Corporation)

5

JOINING

In the construction of electronic equipment, metal parts must be joined together to form rigid bonds. Metals are joined by soldering, brazing, or welding, which are defined as follows:

1. SOLDERING is joining of metals using a *filler material* which has a flow temperature less than 800°F. This is popularly called *soft soldering*.
2. BRAZING is joining of metals using a *filler material* which has a flow temperature greater than 800°F, but less than that of the metals being joined. This is sometimes called *hard soldering* or *silver soldering*.
3. WELDING is joining of metals at or above the melting point of the metals being joined. A filler may or may not be used. If used, the melting point of the filler is approximately the same as that of the metals being joined.

Distortion during joining is a function of temperature. Soldering is used when close tolerances are required. Brazing causes more distortion in the base metal than soldering, and welding causes more than brazing.

As in other operations, an important consideration is room to work in. For joining, this means room for a soldering iron or torch, or proper location of welding electrodes. Precautions must be taken to ensure that the worker is not endangered by drops of molten metal during the joining operation.

5-1 TYPES OF JOINTS

Two pieces of metal may be butted together at an edge and joined by soldering, brazing, or welding. This is called a butt joint and it is the simplest, although not the strongest, method of joining. The butt joint and other basic joints are shown in Fig. 5–1. These are used in frames and other structures.

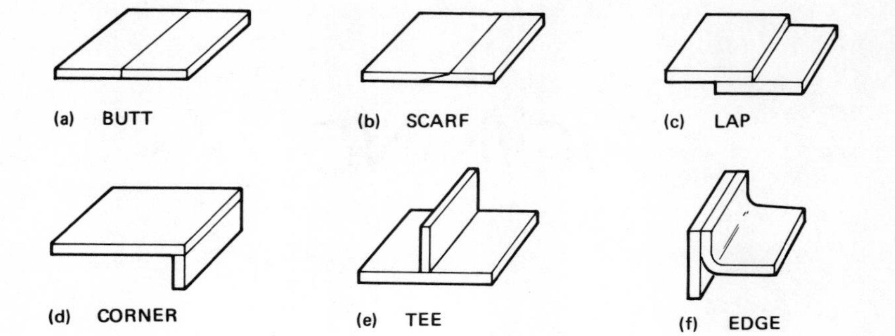

FIGURE 5-1 Basic Joints

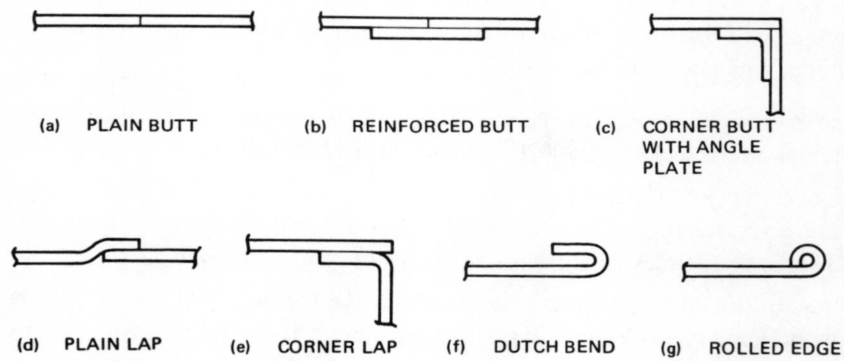

FIGURE 5-2 Sheet Metal Joints

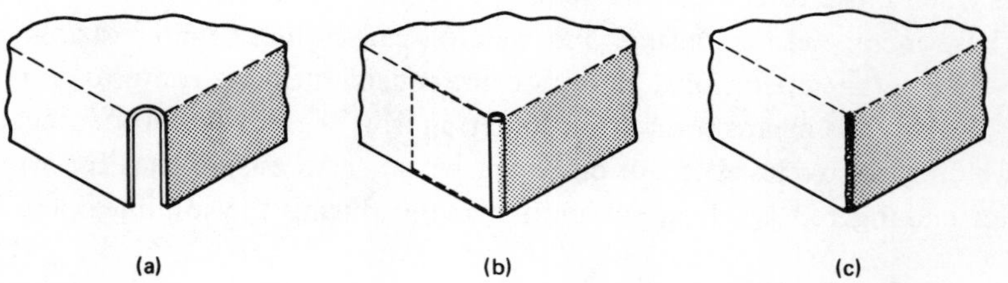

FIGURE 5-3 Corner Joints

Similar joints used in sheetmetal work are shown in Fig. 5–2. Reinforcing plates shown in (b) and (c) give added strength to the joint.

Corner joints, shown in Fig. 5–3, are found in all flanged covers, boxes, and some chassis. The open corner shown in (a) is not usually joined and is acceptable for internal box-like structures where the metal is sufficiently rigid without further stiffening. The lap joint (b) is usually spot-welded. The butt joint (c) may be welded, soldered, or brazed, depending on the metal and the application. When spot welding is used, as in the corner of Fig. 5–3(b) or the lap joint of Fig. 5–1(c), soldering may be used afterward for appearances or for sealing.

5-2 SOLDERING

In soldering, the base metal is heated to a temperature higher than the melting point of the filler material but less than that of the base metal. Some common methods of applying the heat are soldering iron, torch, oven, induction coil, and hot bath. The most popular method is soldering with an electric soldering iron which has a copper point. In *torch soldering*, a gas flame is substituted for the iron. For *oven soldering*, the parts are placed in a jig so that their relative positions are maintained. Solder is placed in the proper places, then the whole assembly is placed in an oven or furnace. In *induction soldering*, the heat necessary to melt the solder is induced by current in an induction coil. In *dip soldering*, the parts are immersed in a bath of hot solder.

The filler material, that is, the solder itself, is an alloy of lead and tin. It is usually specified by the percentages of the two metals, with the percentage of lead given first. Thus 80/20 solder has 80% lead and 20% tin.

The melting point of the solder depends on the percentages of the two metals. The lowest melting point occurs in 38/62 which is called *eutectic solder*. This solder, with 38% lead and 62% tin, melts at 350°F and also resolidifies at the same temperature. It is preferred if this low melting point is necessary to prevent heat injury to adjacent parts. However, since tin is a relatively expensive metal, solders using less tin are more popular. A commonly used mixture is 50/50. Eutectic solder, as noted above, melts when it is heated up to 350°F and solidifies when it is cooled from a higher temperature down to 350°F. All other mixtures have different melting points and solidifying or freezing points. Thus, the popular 50/50 melts when it is heated up to 420°F, but doesn't solidify until it is cooled down to 360°F. In the range between these two temperatures it has a mushy consistency.

The tensile strength and shear strength of the solder depend upon the percentages of lead and tin. Values for several mixtures are shown in Table 5-1. The strongest is eutectic solder.

Soldering is used for electrical connections and in other applications where there are no mechanical or structural requirements. Since solder joints are weaker than the base metals, parts should be joined mechanically before soldering. The solder itself must flow into the joint; therefore there should be clearance of 0.002 to 0.005 in between the parts or surfaces being soldered.

Tin and silver are easily soldered. Copper, nickel, steel, and Monel tend to oxidize, but they, too, can be soldered if a flux is used to break down the oxides on the surface. Acid flux should never be used to solder wires or parts of circuits, since electric current tends to break down acid joints. Aluminum and magnesium cannot readily be soldered unless they are first plated with zinc or tin. Plating with tin or silver is frequently used on copper or steel to improve solderability.

5-3 BRAZING

Bonding in the brazing process results from alloying and diffusion of the filler metal with the base metal. A brazed bond is as strong as the base metal, although there is some distortion due to heating. If two dissimilar metals are brazed together, there may be more distortion due to unequal thermal coefficients of expansion.

The methods of applying heat are similar to those used for soldering. In *torch brazing*, the joint is heated by a flame from a gas torch. Filler is usually applied after the joint reaches the proper temperature, although it is sometimes placed in position earlier. *Furnace brazing*, like oven soldering, requires that the parts and filler be assembled in a jig and then placed in a furnace. This method is ideal for production since the temperature is accurately controlled, and many assemblies may be brazed simultaneously. *Induction brazing* uses an induction coil and is similar to induction soldering. *Dip brazing* is similar to dip soldering. The jigged assembly, without filler metal, is immersed in a bath of molten filler metal. This method is used mainly for brazing connections and small parts.

Although the temperatures used for brazing are below the melting points of the base metals, they are sufficiently high to affect the properties of the metals. If the mechanical properties of the base metal were obtained by thermal treatment, the high brazing temperature may affect these properties. Work-hardened materials are affected also, since the brazing temperature may be above the annealing temperature and thus will cause the material to soften. Base metals which are already in an annealed state generally will not be affected by the brazing temperatures.

The brazing characteristics of aluminum alloys, coppers, and steels are shown in Tables 3–15, 3–16, and 3–17, respectively. In addition, aluminums which can be brazed to aluminums may also be brazed to low carbon steels. Magnesium and its alloys may be brazed only to magnesium. Low carbon steels, stainless steels, copper alloys except leaded brass, and nickel alloys may be brazed to each other as well as to themselves.

5-4 WELDING

In the welding process, the base metal is melted. If two different metals are used, the temperature must be higher than the higher melting point. However, since there is no upper limit on the temperature as there is in brazing, the temperature is usually much higher than the melting point of the base metal. There are four standard welding techniques:

1. *Oxyacetylene welding*. An oxyacetylene flame produces a temperature of about 5800°F, which will melt the metals most commonly welded. During the process both base metals and filler rod are melted into a common puddle which forms the bond when it cools.
2. *Arc welding*. An electric current in the form of an arc is passed between the filler electrode and the work to be welded. The temperature produced exceeds 10,000°F. Both the electrode and the base metal melt and puddle together. At these high temperatures, the oxygen and nitrogen in the atmosphere tend to combine with the metals and weaken the bond. Consequently, the electrode is covered with a flux which also melts but floats to the surface of the molten puddle, thus protecting it from atmospheric elements.
3. *Inert gas arc welding*. This is the same as arc welding, except that the welding is done in a controlled atmosphere of inert gas which protects the bond from oxidation and nitrogenation. No flux is required.
4. *Spot welding or resistance welding*. This method does not use a filler metal and is the most widely used welding technique. A pair of electrodes are clamped to opposite sides of the joint and a high current is passed through. Heat is produced by I^2R loss in the metal, and the metal melts. The parts are pressed together and the current removed, resulting in a bond at the point of contact. Seam welding consists of making a continuous or almost continuous series of spot welds, using a rolling wheel as an electrode and making and breaking the current as required.

The weldability of various metals was shown in Tables 3–15, 3–16 and 3–17. In addition some of these metals may be welded to other metals. Table 5–2 shows the weldability of metals to other metals by spot welding. In this table, weldability is rated from A to D, representing excellent, good, fair and poor weldability. The letter X indicates not weldable.

The shear strength of a spot weld bond depends on the thickness and strength of the material which is welded and to some degree on the number and spacing of the spots. The *maximum* design shear strength in pounds per spot is given in Table 5–3 as a function of thickness for some aluminum alloys. These values may be used in design since they already have a safety factor built-in. Individual spots may in fact be stronger.

In spot welding, the designer must specify the number of welds and their locations. Welds cannot be too close together. After a single weld has been made, this bond presents a low resistance path for the current. The next spot must be far enough away so that this low resistance path does not shunt the current from the location for the next weld. The spot weld creates a certain amount of distortion and wrinkling in the metal, and in order to have the edge undistorted, the spot must be kept away from the edge. These distances depend on the thickness of the material. Table 5–4 lists minimum overlap required of aluminum pieces to be bonded, spot spacing, and edge distance for spot welds in aluminum. These are preferred design values. Table 5–5 lists the same data for low carbon steel, as well as minimum shear strength in pounds per spot. Spot spacing may be increased to four or five times the values in Tables 5–4 and 5–5, if the additional shear strength is not needed.

Welding can be used for joints that must withstand heat which would melt a brazed joint. Welded joints have the strength of the annealed base metal. For best joints, pieces to be welded should have the same or nearly the same melting temperatures and approximately the same thickness.

Although magnesium is not a recommended material for electronic assemblies, it is sometimes required because of its light weight. It can be spot welded, and dimensions for spot welding magnesium are given in Table 5–6. Minimum overlap is the same as for aluminum.

5-5 RIVETS

Riveting is not really joining, since parts are fastened together rather than being bonded. However, it is included in this chapter, since riveting and bonding serve the same purpose. Riveting is more expensive and slower than spot welding and therefore is not used if spot welding can be used instead. However, it does have some advantages. It is used whenever heat distortion from spot welding may be a problem, or when welding is impossible. For example, riveting may be used to join non-metallic materials to each other, non-metallic material to metal, or two metals with widely different melting points. Rivets may be made of aluminums, steels, coppers, and other alloys.

There are three basic types of rivets. *Solid rivets* are used in frames and other structures which may carry relatively heavy loads. *Tubular rivets* require less force to form them and thus are used to join materials which are too soft or too fragile to withstand the force necessary to form solid rivets. *Blind rivets* are used where one side of the joint is inaccessible.

Rivets have been made with a variety of heads, including cone head, button, round top, flat top countersunk, flat head, and many others. The two most popular and preferred styles are shown in Fig. 5–4. The *universal head* is used in all cases where the projecting head is not bothersome. The *100° countersunk head* has a flat top and is used where a flush surface is needed. The protruding shank of the rivet is turned over to form another head, called the *upset head*, and this head also may be rounded like the universal head or countersunk for flush mounting. In general, countersunk heads should be avoided, since the required countersinking is an extra operation which adds to the cost of production.

The *length* of the rivet is the length of the shank in a rivet with a universal head, and it is the overall length in a rivet with a countersunk head. When

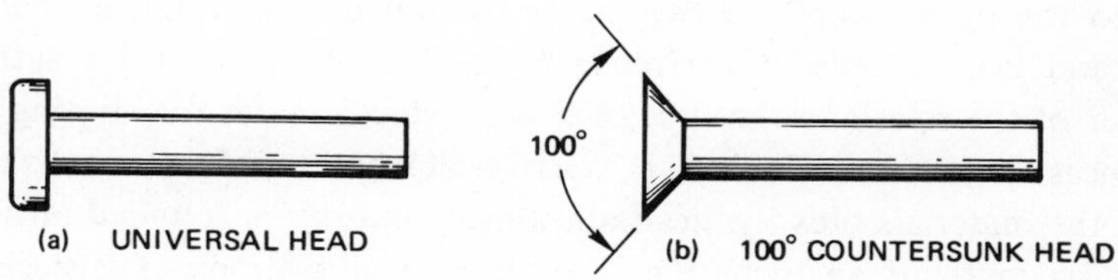

FIGURE 5-4 Preferred Rivets

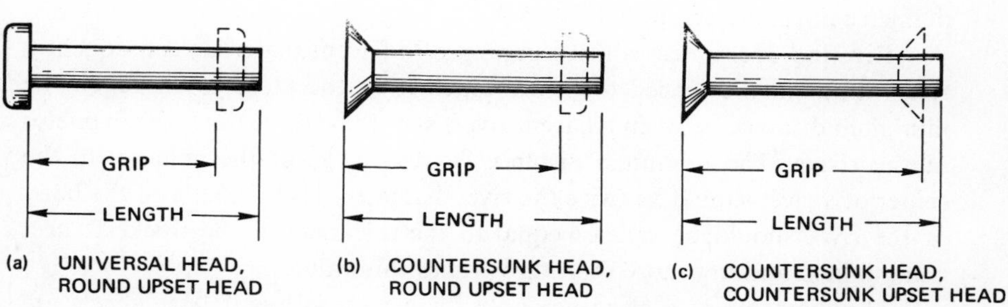

FIGURE 5-5 Rivet Dimensions

the rivet is fastened, the *grip* is the spacing between the flat portion of the two heads and it is equal to the total thickness of materials riveted together. These dimensions are shown in Fig. 5–5.

The rivet material must be similar to the metals being joined to prevent corrosion caused by contact of dissimilar metals. This does not apply to non-metallic materials. The popular aluminum rivets are made of one of three alloys. Alloy 1100–2S, with a shear strength of 11,000 pounds per square inch, is preferred. Where more strength is required, 2117–T4 with a shear strength of 33,000 PSI, or 2017–T4 with a shear strength of 35,000 PSI may be used. Rivets made of cadmium-plated steel or Monel are used to rivet steels. These rivets have a shear strength of about 40,000 PSI.

Solid rivets are called out by a style number, material designation, diameter, and length. Style numbers for aluminum solid rivets are MS20426 for the style with the countersunk head and MS20470 for the universal head. The alloys are designated A for 1100–2S, AD for 2117–T4, and D for 2017–T4. Body diameter of a rivet is given in thirty-seconds of an inch and the length in sixteenths of an inch. For example, MS20426A4–6 would indicate a countersunk rivet made of 1100–2S aluminum, $\frac{4}{32}$ of an inch in diameter and $\frac{6}{16}$ of an inch long.

The length of a solid rivet depends on the thickness of the materials being joined and on the diameter of the rivets used. In Fig. 5–5, the grip is equal to the thickness of the two materials joined. The difference between length and grip is called the *riveting head allowance* and varies with the diameter of the rivet and the type of head. Typical values for riveting head allowances are shown in Table 5–7. The overall length is, of course, the thickness of the materials plus the head allowance, and this is rounded off to the nearest sixteenth of an inch. For clearance considerations, it is necessary to know the maximum size of the round upset head and how far it protrudes above the surface. The maximum diameters and maximum heights for various diameter rivets are given in Table 5–8.

Riveting does cause some crimping of adjacent material; consequently rivets must not be spaced too close together or too close to an edge. The minimum distance between adjacent rivets should be three times the diameter of the rivets. The minimum distance from an edge of the material to the center of a rivet should be twice the rivet diameter. The diameter of the body of the rivet should be at least equal to the thickness of the thickest sheet being joined and not more than three times this value.

The diameter of a clearance hole for a rivet is less if both sheets are drilled simultaneously than if the holes are drilled in the two sheets separately.

If the holes are drilled in one sheet, then transferred to the second, the same hole size used in simultaneous drilling may be selected. Clearance hole diameters for simultaneous drilling or transferred holes are given in Table 5–9. For non-transferred holes, the diameters should be increased by about 0.005 in.

6

MACHINING

The designer furnishes only the shape and dimensions of the parts to be machined, but does not specify how the parts are to be made. Nevertheless, the designer must be familiar with all machining techniques, because the design usually dictates which technique will be used. Designs should be chosen which can be built with minimum handling in order to minimize cost. On the other hand, close tolerances on dimensions require more care, and sometimes it is better to select a design which requires an extra set-up but permits looser tolerances. If nothing else, the reject rate will be reduced.

6-1 DRILLING

The drill press is used to drill round holes. In sheet metal, round holes are usually punched, unless the diameter of the hole is less than the thickness of the material, but in other parts of the system holes are usually drilled. Where holes are indicated in the design, there must be clearance for the drill to reach the point where the hole is to be drilled and to pass through the material. Extra long drill bits are available, but they are special tools which increase the cost of the job and may delay the work.

The drill should enter the material perpendicular to the surface. If a hole is needed on a curved surface or at an acute angle to a flat surface, it is desirable to use a flat boss or to mill a flat on the surface perpendicular to the drill direction. These techniques are illustrated in Figs. 6–1 and 6–2.

In Fig. 6–1(a), without a boss, the drill will tend to bend and slide down the surface. The addition of the boss, as shown in Fig. (b), prevents this. When drilling at an angle, as shown in Fig. 6–2(a), the drill also tends to slide or wander. This is prevented by a milled flat as shown in (b). The milled flat could also be used on the curved surface and the boss for the angle drilling as well.

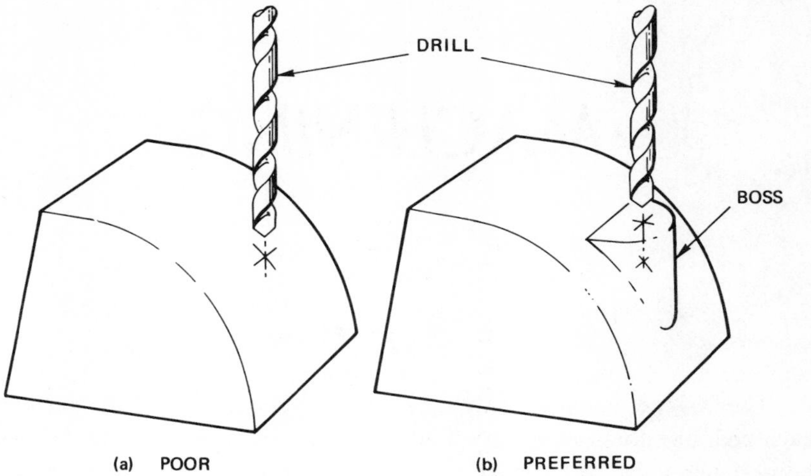

FIGURE 6-1 Drilling on Curved Surface

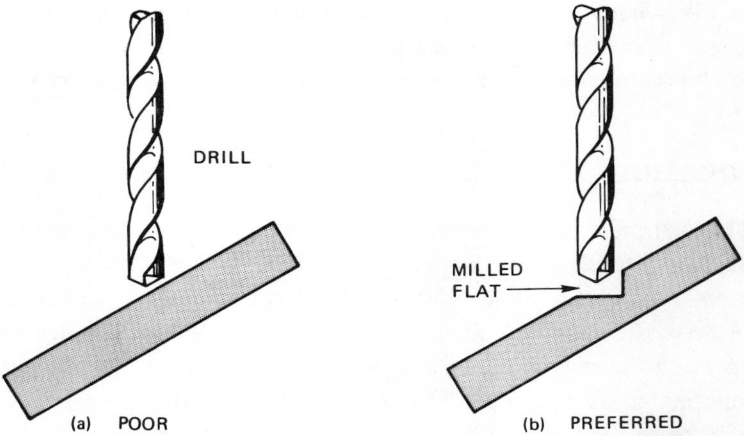

FIGURE 6-2 Drilling at an Angle

When two parts have mating holes, it is possible to drill through both at once and thus be sure they match accurately. This is called *matched drilling*. In general, tolerances on hole locations and hole sizes may be relaxed somewhat when the holes are drilled simultaneously, and this may be an advantage. Nevertheless, matched drilling should be avoided in general, since it requires that the two parts to be drilled must be brought to the drill at the same time. If other processes which differ are required on the two parts, it may be necessary to delay one process in order to drill them together.

However, matched drilling may be specified if it represents a definite savings in time or relaxation of tolerances (which also saves time).

Every shop should have a complete set of standard drills, and, of course, standard drill sizes should be specified. Standard sizes of both numbered drills and fractional drills are shown in Table 6–1. Normal tolerances on hole sizes are shown in Table 6–2. These are tolerances which are easily met. Closer tolerances are achievable, but require more time and thus are more expensive. If possible, the loose tolerances shown in Table 6–2 should be used.

Although all drill sizes are readily available, the number of *different* sizes used on any one job should be held to a minimum. Each different size represents an additional set-up and therefore requires additional time. This is especially important when a multi-spindle press is available. The number of different hole sizes should not exceed the number of spindles.

6-2 TURNING

A lathe is used to produce circularly symmetrical parts such as bushings, sleeves, discs, rods, and axles. Axial holes and internal and external threads are also made on a lathe. However, the lathe is not usually used for production of large quantities of a single part. For production, it is more desirable to use extrusions or castings, although a lathe may be used if extremely high accuracy is necessary.

In general, centers are not shown on the design drawing, but they may be permitted on the part. If centers are *not* permitted, it should be so stated by a note on the drawing. Acceptable notes are indicated in Fig. 6–3. Centers

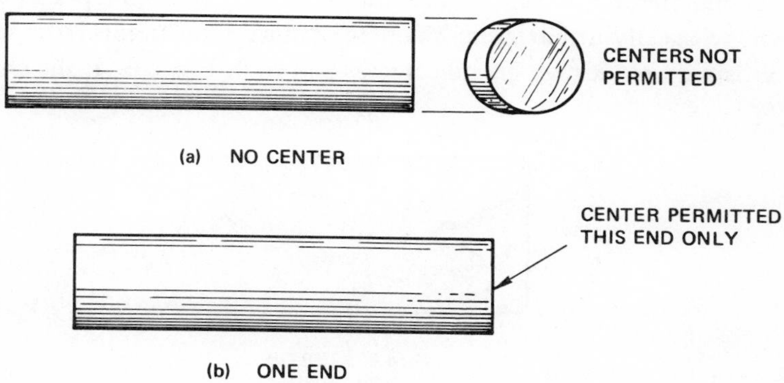

FIGURE 6-3 Center Restrictions

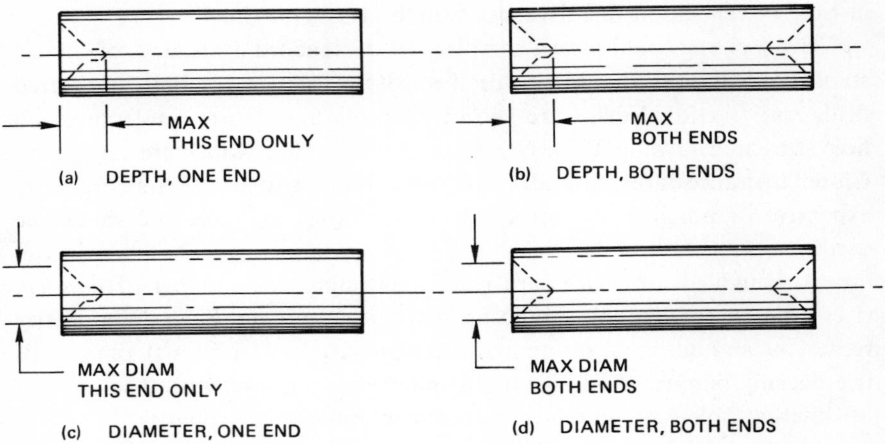

FIGURE 6-4 Maximum Dimensions of Center Holes

may be permitted as long as they stay within certain maximum limits of diameter or length. The restriction may apply to one or both ends. Possible configurations are shown in Fig. 6–4.

When a part requires a center for a functional purpose, it is shown as indicated in the example of Fig. 6–5. It should be noted that the depth is usually omitted, unless it is the important dimension, in which case the diameter is omitted. In the example of Fig. 6–5, a #4 center drill is shown. The sizes of standard center drills are shown in Table 6–3. The length of the small diameter is equal to this diameter. All of these drills have a 60° included angle between the two diameters.

It is usual to break the sharp edge of an external diameter on a rod by a radius or chamfer. If the chamfer or radius serves no functional purposes, it is sufficient to specify a maximum chamfer or maximum radius. If a specific chamfer is required it may be indicated as shown in Fig. 6–6. If the chamfer

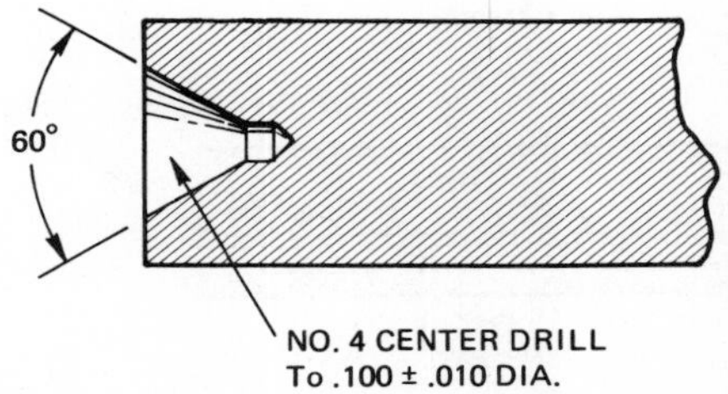

FIGURE 6-5 Functional Center

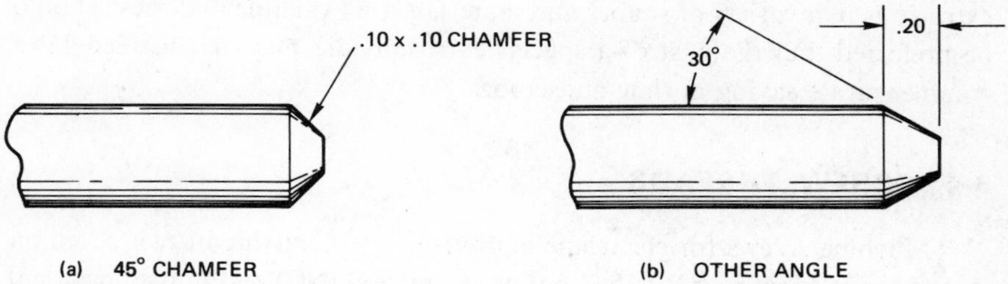

FIGURE 6-6 Chamfers

is at 45°, it is shown by its dimensions, as in (a). For an angle other than 45°, the angle is given as shown in (b).

Countersunk holes are used when screw heads and rivet heads must not protrude beyond the surface. If possible, a countersink should be avoided in design since it requires an extra operation. In sheet metal, countersinks are produced by punching or coining, but in solid metal parts, they are turned on the lathe or infrequently on the drill press. For countersunk rivets, the included angle is 100° ± 2°. Other standard countersink sizes contain included angles of 82°, 90°, and 120°, and these usually have a tolerance of ±10°.

6-3 MILLING

A milling machine is used to form non-round holes and openings, keyways, and other irregular shapes. It is also used to form flat surfaces and curved surfaces. On the lathe, the part to be worked rotates, and a single-toothed cutting tool is applied to it. In contrast, on the milling machine, a multi-toothed cutter is rotated, and the work is moved. Standard milling cutters are available for a variety of tasks, and the design should permit use of these standard cutters. Special tools can be made, but at considerable additional expense and delay.

Internal openings and holes are made on a milling machine. Internal square corners should be avoided since they are more difficult to machine and are structurally weaker than rounded corners. Wherever possible, a radius should be indicated, either as a specific dimension for a functional purpose or as a maximum permitted dimension to simplify the milling operation.

Parts which will require milling must be rigid enough to be clamped to the table of the milling machine and to withstand the action of the cutting tool without bending. In many cases extrusions may be preferable. Standard extrusions in a variety of spapes and materials are available and these should be preferred, but the cost of a special extrusion die may be justified by a commensurate saving in time and labor.

6-4 SCREW THREADS

Machine screws for electronic equipment are available in two common thread styles, designated unified national coarse (UNC) and unified national fine (UNF). The U is sometimes omitted, and the threads are designated simply NC or NF. Extra fine threads are also available but are rarely used for electrical or electronic gear. Screw sizes are usually specified by the diameter, or a number indicating the diameter, followed by the number of threads per inch. For diameters under $\frac{1}{4}$ in. numbers are used to designate the screw size, and actual diameters are used from $\frac{1}{4}$ in. upwards. A size zero screw is 0.060 in diameter and each higher number represents a 0.013 in. increase in diameter. Thus, the actual diameter for any screw size is given by

$$D = 13N + 60 \tag{6-1}$$

where D is the diameter in thousandths of an inch and N is the screw size. For example, a #10 screw has a diameter of 0.190 in.

Standard screw sizes for both NC and NF threads are shown in the first two columns of Table 6–4. The third column gives the diameter of the screw or tap. For the numbered screw, the values in this column may be obtained from Eq. 6–1. Also shown in this table are tap drills and clearance drills for the different screw sizes.

Since the difference in diameter between adjacent screw sizes is only 0.013 in., there is not much difference in the strength of screws of adjacent sizes. For all practical designs, fewer different sizes would be sufficient, and consequently most companies prefer to stock a minimum number of sizes which become their standard or preferred sizes. A representative list of preferred sizes might be #2-56, #4-40, #6-32, #8-32, #10-32, and $\frac{1}{4}$-28. The sizes of screws used in a design should, if possible, comply with the company's standard list, but—more important—the number of different sizes or types of screws (or in fact any hardware) on any one job should be held to a minimum to simplify assembly procedures.

Blind holes should be avoided since an extra step is required in the manufacturing process to make sure the depth of the hole is correct. If tapped blind holes are necessary, the length of the hole must include an allowance to catch the chips formed while tapping.

In soft materials such as aluminum, magnesium, and some plastics, screws which are frequently removed will eventually Strip the threads in the holes. To prevent this, helical threaded inserts made of corrosion resistant steel are used. The coil is wound into a tapped hole in the metal and the internal diameter of the coil forms a unified coarse or fine thread, as desired. Data for sizes of inserts, hole diameters, and special taps are available from the manufacturers.

6-5 NUMERICAL CONTROL (NC)

Automation is a method of eliminating human errors during production. Once an automated assembly line or automated machine shop has been set up to perform a task, it will work indefinitely without making a mistake or getting tired. In quantity the cost per piece may also be reduced. For a single piece it does not pay to use an automated set-up, but, depending on the type of automation, the break-even point may occur with relatively few pieces.

Numerically controlled equipment operates from data on a punched card or tape. Data are entered on punched tape or magnetic tape in the form of a binary code, and this information moves the tables, spindles, punches, etc., to the proper positions for each operation.

The accuracy of a numerically controlled machine depends on the accuracy of the program. It is possible to maintain positioning accuracy of less than a thousandth of an inch, and three thousandths is routine. The program may be prepared by a programmer directly from the drawings. On some machines, a machinist may make the first piece, and his operations may be transferred automatically to a tape. When the tape is fed back into the machine, it performs the same operations that were done by the machinist.

Writing the program and preparing the tape can usually be done in less time than it would require to make the jigs and fixtures necessary to do the job by conventional machining. No jigs are needed on a numerically controlled machine, since all points are simply referred to two reference lines. Although it may not pay to have a numerically controlled machine in a shop where only short runs are performed, once a program is written, it is economical to use numerical control even for a single piece. Thus, if 100 parts are made and it is found that a few more are needed, the NC machine can be set up to do the additional parts more economically than they could be done conventionally.

In general, a designer is not concerned with the types of machines in the shops. However, if he knows that numerically controlled machines will

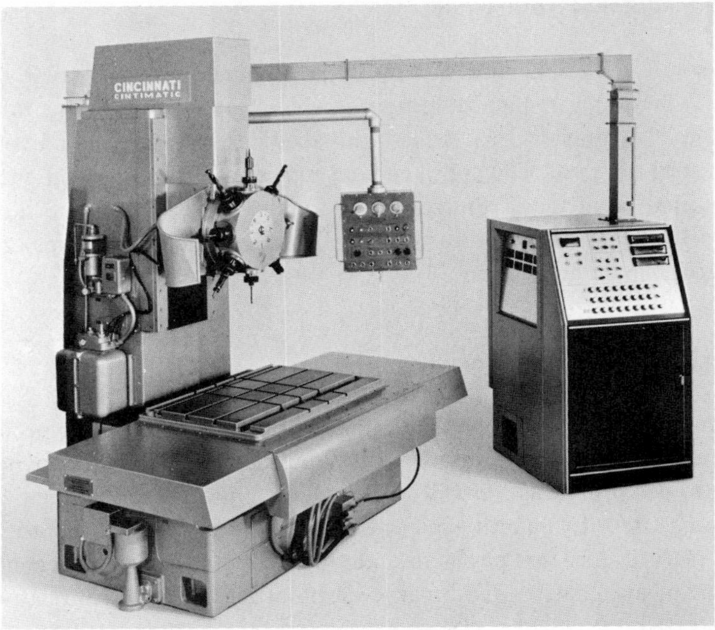

FIGURE 6-7 Eight-Spindle Turret Drill (Photo Courtesy of Cincinnati Milacron.)

FIGURE 6-8 Numerical Controlled Machining Center (Courtesy of Monarch Machine Tool Co.)

be used, there are steps he can take to simplify the programmer's job. All distances in the program are measured from two reference lines, usually the bottom edge and left-hand edge of the work as it rests on the bed of the machine. If angular dimensions are normally used in the drawing, the designer should add x and y coordinates as reference dimensions. Dimensions should be given as distances from the reference lines whenever possible.

Numerical control is available for practically all types of machines. Representative NC machines are shown in Figs. 6–7 and 6–8. Figure 6–7 illustrates an 8-spindle turret drill. The control console is shown at right. In operation the turret rotates automatically so that the proper drill, countersink, reamer, or other rotating tool is in position. The table moves to the correct x and y coordinates, and the tool moves down. All this is controlled by the binary code on the tape. In designing for this machine, it is important that not more than eight different tools be required to do the job. For a shop with only a 6-spindle drill, only six tools should be required. If a single additional tool is needed, it means an extra tape and an extra set-up.

The numerically controlled machining center shown in Fig. 6–8 can do milling, punching, and other tasks. A variety of tools is available, and the proper tool is brought to the operating position at a signal from the tape in the console at the right of the picture. The material to be worked on is also moved automatically to the proper position. Again, the design should permit machining in one set-up. Whenever NC machines are available, the designer must be aware of their capabilities, including the number and kinds of operations they can perform with a single tape.

6-6 DIMENSIONING AND TOLERANCES

The most important consideration in dimensioning a drawing is how best to facilitate the machinist's task. The designer should understand and visualize how the machinist will go about making the part and also where the possible sources of careless errors may be. Although the fundamental rules of drawing dictate that no point or line should be located by more than one dimension, the designer should add a note or reference dimension whenever it will help avoid a careless error. All horizontal dimensions, notes and other text should be positioned to be read when the drawing is in a normal position, with the bottom edge toward the reader. All vertical dimensions should be placed to be read in one direction only, preferably with the right-hand edge toward the reader.

Rectangular dimensions should be used in all cases, unless an angular dimension is critical. Even in this case, the rectangular dimensions should be referenced. This is especially important for numerical control.

The drawing must contain sufficient views and sections to show all necessary dimensions. Every line, hole, point, etc, must be explained clearly so that no scaling is required and no assumptions are necessary. A dimension locating a position or indicating a size must have a tolerance showing the limits of possible deviation. Reference dimensions do not have tolerances.

Functional dimensioning is illustrated in Fig. 6–9 and indicates how a designer can help a machinist. When using a milling machine, a machinist locates the first hole from the edges of the material. This is hole A. Then he moves the table to hole B and finally to hole C. In each case he wants to know the amount of movement. This is presented to him in the drawing of Fig. 6–9(a). If the same job is to be done on a numerically controlled milling machine, the programmer would want to know all locations with respect to the reference edges. This is presented in (b) for the same hole locations. No tolerances are indicated in Fig. 6–9. However, where tolerances on positions are critical, the arrangement of (b) is preferable, since in (a) tolerances are cumulative.

Tolerances are usually given for every dimension. However, a box which indicates tolerances to be used may be included, then these tolerances apply to every dimension which is not otherwise toleranced. Tight tolerances are expensive. In general, the cost of achieving a tolerance of ± 0.001 in. on a milling machine or lathe is more than double the cost for a tolerance of ± 0.005 in. The tolerances on hole sizes in Table 6–2 are easily achievable. To attain better tolerances, it is first necessary to drill a smaller pilot hole and then finish with the correct drill size. This adds an extra operation to each hole.

The *basic* dimension is the dimension desired. Tolerances may be indicated as a permissible deviation in either direction from this basic dimension, as

$$0.500 \pm 0.004$$

This is a *bilateral* tolerance. Tolerances may be indicated as a permissible variation in one direction only, as

$$0.496 \begin{array}{l} + 0.008 \\ - 0.000 \end{array}$$

This is a *unilateral* tolerance. In both cases, the permissible variation is from

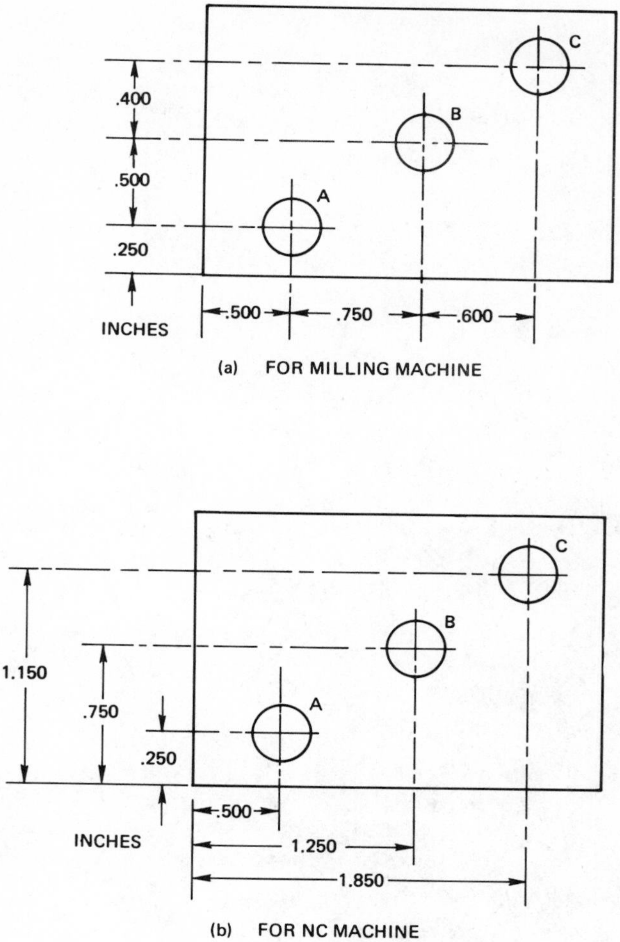

FIGURE 6-9 Hole Positions

0.496 to 0.504. This may also be expressed by simply giving the limits, as

$$\frac{0.504}{0.496} \quad \text{or} \quad \frac{0.496}{0.504}$$

When using limit tolerances, the upper number should be the dimension the machinist aims for so that if he passes it, he is still within the limits. Thus, for an external diameter the upper number would be the larger limit, and for an internal diameter it would be the lower limit.

7

FINISHES

Metallic surfaces in electronic equipment usually require some sort of treatment during or after fabrication for protection, appearance, or electric or mechanical reasons. Surface finishes may be categorized in a number of ways. They may be organic or inorganic. They may be applied chemically, electrically, or physically. They may be specified by function. There are four general purposes for surface finishes. The most important is protection from contaminants and corrosion. Others are conduction, solderability, and appearance. Usually one finish may serve two or more functions.

7-1 GALVANIC CORROSION

When two dissimilar metals are placed in a solution which conducts electricity, they form a battery. That is, there is a potential difference between the ends protruding from the conducting solution, and if these ends are connected a current will flow. A typical electrochemical cell is shown in Fig. 7–1. The conducting solution is called *electrolyte*, and the cell acts as a battery. A potential difference exists between the ends of the copper and the zinc strips. If the ends are connected, current flows. That is, electrons flow from the zinc to the copper. The result is a dissolution of the negative pole, in this case the zinc.

If two dissimilar metals are in contact and moisture gets between them, the same type of electrochemical cell is produced accidentally. In general, the metals in contact are not perfectly smooth. Moisture collects in the spaces, producing a battery, and because the metals actually make contact at some points, current flows. Again there is dissolution of the negative electrode, this time unwanted and unexpected. This is called *galvanic corrosion* and is illustrated in Fig. 7–2.

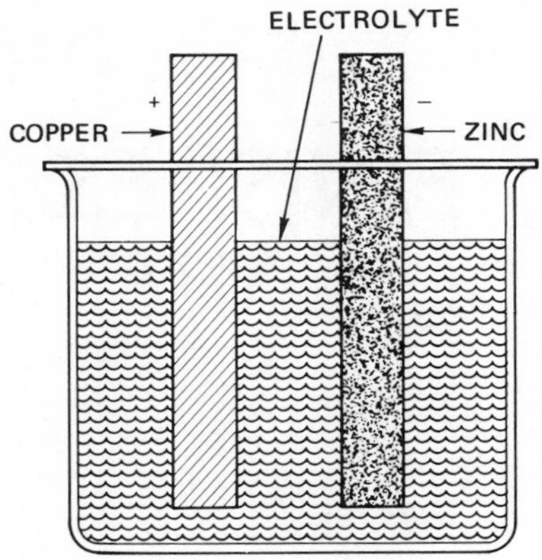

FIGURE 7-1 Electrochemical Cell

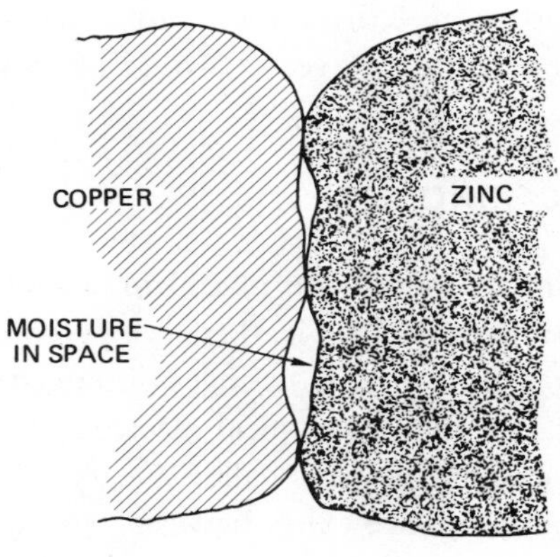

FIGURE 7-2 Galvanic Corrosion

The potential which any metal will attain with respect to gold is called the *anodic index* for that metal and is measured in hundredths of a volt. A list of metals arranged according to anodic indices is called a *galvanic series*, as shown in Table 7–1. The values for the anodic indices in this table are approximate and depend on the degree of impurity or the kinds of additives in the alloys. From the table, a cell containing copper and gold should have a potential difference of 0.35 volt. The potential difference created for any two metals is the difference between their anodic indices. Thus, a copper-zinc cell should have a potential difference of 0.90 volt (125–35). In the galvanic series, as listed in Table 7–1, the upper end is called the *cathodic end,* and the lower is called the *anodic end,* Since the metal toward the anodic end is the one which corrodes when two are brought together, the anodic end is also called the *corroded end,* while the cathodic end is called the *protected end.*

The amount of corrosion that may occur when dissimilar metals are in contact is a function of the potential difference between the two metals. If this difference is 0.25 volt or less, the galvanic action should be negligible. Thus, there would be no danger in placing silver and copper (0.20 volt difference) in contact. The higher the difference in potential, the greater is the likelihood for corrosion, and the worse will be the corrosion when it occurs. Table 7–1 points up a major disadvantage of using magnesium, since it has a high potential difference with all other metals and is thus more susceptible to galvanic corrosion.

Galvanic corrosion is not a serious threat since it is possible to prevent it or minimize it by proper design. Wherever possible, compatible metals should be used when contact must take place. However, even where dissimilar metals are necessary, plating can solve the problem. Thus, if aluminum and copper must be placed in contact, the aluminum can be copper-plated. There is no galvanic action between a base metal and a plating when properly done, since no moisture can get between the two. If electrical contact is not essential, corrosion may be prevented by using insulation between the two metals. This can be in the form of washers or by insulating surface coatings. A sometimes neglected area is the use of screws and nuts which are of a dissimilar metal from one or both of the parts being joined. Proper plating on the hardware should be considered to prevent galvanic corrosion.

7-2 ENVIRONMENTAL CORROSION

The atmosphere frequently contains a number of pollutants which attack metals chemically. The degree of impurities in the air ranges from the relatively clean rural environment to the moist, salt air of a marine environment. Moisture can be damaging. Moisture with chemical agents added, such as salt, is especially corrosive. For convenience, atmospheric conditions are grouped into three categories. *Normal* is applied typically to an uncontaminated, rural atmosphere. *Industrial* is applied to a contaminated, urban atmosphere. *Marine* indicates a moist atmosphere containing contaminants such as salt.

When metal is exposed to a corrosive atmosphere, it tends to corrode first and most rapidly at points of unusual stress. This can be minimized by avoiding sharp corners, bends, and undue strains in the metal. Exposure to moisture or contaminants causes the surface of the metal to dissolve at these points of stress. Moisture then gets into the cracks thus formed and causes corrosion deep into the metal.

The *corrosion resistance* of a metal is its ability to withstand environmental corrosion. *Pure* aluminum is unusually resistant to corrosion, because on contact with air, a hard protective oxide coating is formed. This coating does not flake but remains on the surface, effectively preventing corrosive elements in the atmosphere from reaching the metal. In contrast the oxidation on steel is flaky and falls off, exposing the surface to further corrosion. The corrosion resistance of aluminum alloys depends to a large extent on the additive in the alloys, but also on the degree of stress relieving and temper. Table 7–2 indicates the relative corrosion resistance of the popular aluminum alloys to the three types of atmospheres. From this table it is evident that clad metals are better than the same metal unclad, and indeed this is the main reason for cladding. Alloy 5052, which contains magnesium, is especially resistant to all varieties of environmental corrosion and should be selected wherever a marine atmosphere will be encountered.

7-3 CORROSION PREVENTION

Prevention of environmental corrosion must be considered in designing electronic equipment, since corrosion can cause electrical failure as well as structural defects. Equipment which is to be used on shipboard must be "moisture-proofed" more thoroughly than a piece of equipment that is to be used in an air-conditioned, controlled atmosphere, or even in a home. Thus, a marine radar must be designed to be resistant to corrosion, whereas a home television set does not.

It was pointed out in Sec. 7-2 that pure aluminum is especially resistant to corrosion because of the hard oxide that forms on its surface and adheres to the surface, preventing moisture from reaching the metal. In contrast, copper and steel also oxidize, but their oxides flake off leaving the metal exposed for further deterioration. A possible method of preventing corrosion is to cover the metal with a hard, non-corroding surface which will adhere to the base metal and exclude harmful moisture. This indeed is what is done. A protective coating may be applied chemically, as an oxide; electrochemically, as plating; or mechanically, as paint.

Aluminum alloys are generally resistant to corrosion, except for alloys in the 2000 series, as was discussed in Sec. 3-1. However, to enhance corrosion resistance all aluminum alloys may be chemically treated to form an oxide on the surfaces. Oxides may also be deposited electrolytically. There are many patented processes available for doing this, including Alumilite (Alcoa), Alodine (American Chemical Paint Co.), Iridite (Allied Research Products), Bonderite (Parker Rustproof Co.), and others.

It is customary for a company to select one of these processes as its standard, and this should be the one called out on the drawing. It should be noted that the oxide is usually a very thin coating, usually less than 0.0005 in. thick. Consequently, although the oxide coating protects the metal from environmental corrosion, it is not protection against galvanic corrosion when the aluminum is in contact with a dissimilar metal. An additional coat of paint, however, is sufficient protection.

Copper alloys, zinc alloys, and ferrous metals cannot be protected by oxides. Instead they are protected by plating with metals which are corrosion resistant. Common protective platings include nickel, cadmium, and chromium. These three are widely separated in the galvanic series (see Table 7–1), and thus the choice might be influenced by the presence of another metal in contact with the plated metal. Then the plating would prevent galvanic corrosion as well as environmental corrosion.

7-4 ELECTROPLATING

Plating was mentioned in Sec. 7-3 as a method of preventing both environmental and galvanic corrosion. Plating has other uses, also, and is an important surface finish. There are special design considerations for parts which are to be plated.

Plating may be used as a finishing coat. Thus nickel plating may be used as a decorative finish on both copper alloys and ferrous metals. It may be finished to a bright, shiny surface or to a soft, matte finish. Chromium plating is used when a hard surface is needed to resist excessive wear. If corrosion resistance is needed as well as wear resistance, the metal is first nickel plated and then chromium plated. The chromium plating should be about 0.002 in. thick. Nickel plate on copper and copper alloys varies from 0.0001 to 0.0005 in. thick, either as a finish or when chromium is plated over it. On ferrous metals which require more corrosion protection, the nickel plating should be about 0.002 in. thick.

Carbon steel and most copper alloys oxidize quickly, and since the oxides are flaky, surfaces on these metals will not hold paint. When these metals are to be painted, one possible solution to the adhesion problem is to plate the metal with tin to a thickness of 0.0001 to 0.0005 in. Tin plating is also used when these metals are to be soldered, since the oxides prevent adhesion of tin-lead solder to carbon steel or copper.

Plating is also used to improve conduction. This is of importance in the design of equipment to be used at microwave frequencies. At dc and low radio frequencies, current flows through the whole cross section of the metal, so that a thin surface coating will not affect conduction. However, as the frequency is increased, the current moves toward the surface of the metal. This is called the *skin effect*. The current density is greatest at the surface of a conductor and falls off exponentially with depth. The depth at which the current density is 1/e (about 37%) of its value at the surface is called the *skin depth*, and is given by

$$\delta = 2\sqrt{\rho/\mu f} \tag{7-1}$$

where δ is the skin depth in inches
ρ is the resistivity in ohm-centimeters
μ is the permeability (unity for nonmagnetic metals)
f is the frequency in Hertz

For copper, $\rho = 1.724 \times 10^{-6}$, and δ is $2.61/\sqrt{f}$ in. At a frequency of 100 MHz, the skin depth is about one quarter of a thousandth of an inch. At microwaves it is much less and practically all the current flows next to the surface. Thus, to improve conductivity at higher frequencies, the base metal may be plated with copper or silver. For corrosion protection, an additional plating of rhodium may be added. This last coating is less than 0.00001 in. thick, and since rhodium has a higher resistivity, the skin depth is much greater than this. Most of the current then flows in the layer of silver or copper plating which may be less than 0.0005 in. thick.

Metal parts which are to be enclosed in a hermetically sealed compartment normally require no protection from environmental corrosion. However, parts made of ferrous metals tend to rust before assembly and, consequently, should be protected even if they will be hermetically enclosed. The parts may be plated for protection, but the plating should be of a metal which will not react with vapors from insulating materials or lacquers inside the hermetically sealed compartment. Cadmium plating should never be used in sealed compartments for this reason.

Most commonly used metals may be plated. Stainless steel is usually plated first with nickel, since nickel adheres most readily to stainless steel, and other platings adhere easily to nickel. When an underplating is used, for adhesion purposes only, it may be as little as 0.0001 in. thick. Other ferrous metals may be plated directly with copper, silver, and other metals.

Beryllium copper is difficult to plate, but other copper alloys are easily plated. Nickel or nickel and chromium are frequently used. At microwave frequencies, silver and rhodium are used.

Aluminum and magnesium alloys are easily plated. However, plating on pure aluminum may blister under heat. Aluminum alloy 6061 is one of the best for taking and holding plating.

The surface finish of the base metal affects the plating. Castings are especially difficult to plate because of porosity. In general, the finer the finish on the base metal, the better the plating. For microwave cavities, a finish of 8 micro-inches is desirable before plating.

The thickness of plating is a function of electric current. Large, flat areas will tend to have a thinner coat at the center, unless special electrodes are added. Protrusions, edges, and outside corners generally have added plating deposits, whereas inside corners and the bottoms of deep holes have less. Plating in a through hole will be maximum near the ends and taper to a minimum in the middle.

To eliminate uneven plating, sharp corners and protuberances should be avoided in the design. All corners should have a radius of at least 0.015 in. The depth of a hole should not exceed its diameter or width.

Since plating coatings are very thin, they are easily damaged. Even a light nick or scratch can penetrate to the base metal, thus making the plating worthless as corrosion protection. In specifying plating, consideration must be given to how the part is to be handled in fabrication and assembly. Plated parts require extra care.

7-5 ORGANIC FINISHES

Organic finishes include paints, enamels, and lacquers. They are used both for appearance and protection of the surface from the elements. Paint is a liquid containing pigments which dries to a solid film. The pigments provide color and opacity. Enamel is a paint which dries to an extra smooth finish. Enamel is usually baked to dry, but on parts which cannot stand baking temperatures, *air-dry* enamel may be used. Lacquer consists of cellulose nitrate or acetate in a solvent which evaporates quickly, leaving a clear, thin film. It is used to protect parts which cannot be heated and must be dried quickly. It is also used as a transparent, non-tarnish coating for metals and other surfaces. Pigments may also be added to lacquers for coloring.

Organic finishes do not adhere readily to most metals. For improved adhesion, an additional preparation is necessary between the metal and the organic finish. For aluminum, the problem is easily solved, since the chemical oxide used for prevention of corrosion also acts as a good base for paints and enamels. For other metals a zinc-chromate primer may be used. This primer is an effective protection against corrosion itself as well as an undercoat for paint, enamel, or lacquer.

Paint is usually much thicker than the protective oxide coating on aluminum or any plating coating. If dissimilar metals are properly painted, they may be placed together without danger of galvanic corrosion. Proper painting includes cleaning the surface before applying coats.

7-6 SUMMARY OF FINISHES

Large structural parts of carbon steel which cannot be processed in tanks or vats should be covered with a coat of zinc chromate followed by one or more coats of paint.

Smaller parts of carbon steel are cadmium-plated and covered with a coat of zinc chromate. If these parts are not exposed to view in the final assembly, no further finish is needed. If they will be seen, a decorative paint finish may be added.

Carbon steel parts which are in sliding contact where electrical continuity must be maintained are plated with copper first and then nickel. Each plating may range from 0.0003 to 0.0005 in. thick.

Large stainless steel structures should be covered with a coat of zinc chromate followed by a coat of paint. Smaller stainless steel parts, not exposed to view in final assembly, need no finish.

Small copper parts may be plated with nickel or cadmium. If a sliding electrical contact is required, copper alloys are plated with nickel followed by a coating of cadmium.

Structural parts of aluminum are first coated with a chemical oxide treatment, such as Alodine, and then painted with zinc-chromate primer and a final coat of paint. Where parts are not exposed to view, aluminum is simply treated for an oxide coating.

For electrical contact, aluminum is plated first with copper, then with nickel.

Microwave cavities made of aluminum are plated first with copper to about 0.0001 in., then 0.00001 in. of silver and a flash plating of rhodium.

For soldering applications, metals are plated with tin.

8

ASSEMBLY AND WIRING

An electronic *assembly* is a self-contained unit which may contain electronic components, mechanical devices, such as switches, and associated hardware. The assembly itself may be one section in a larger assembly, in which case each of the smaller units are called *subassemblies*.

The process in production of mounting the components and devices on a chassis and connecting them electrically is also called *assembly* or *assembling*. Wiring is a part of the assembly process.

8-1 DESIGN CONSIDERATIONS

In laying out or packaging an assembly, the designer must weigh the restrictions imposed by limitations on cost, size, weight, maintainability, human factors, safety, and operational quality. The final design must be a compromise then in order to achieve in production a piece of equipment which meets electrical specifications and satisfies all other restrictions. The importance of each of the restrictions depends on the market as was discussed at the beginning of this book.

Although simplicity is desirable, a complicated design is not necessarily bad. In general, important considerations are accessibility and safety. The package should be laid out so that there is room to place parts in and on the chassis without requiring special tools or specially trained workers. Parts which may be changed frequently must be accessible so that they can be removed without danger of damaging other parts. Installation of each part must be possible without any risk of damage from heat or tools to parts already installed. An important factor, especially for industrial and military equipment, is that there be room for test-probes used when it is necessary to check the equipment.

Safety must be considered for all people involved in handling the equipment, including assemblers, users, and maintenance men. Sharp corners and edges should be avoided, as should any protrusion which may cause injury or damage to clothing. During assembly, the worker must not have to place his hand near sharp, internal projections. If hand soldering is required, it is important that the worker is in no danger from hot drops of solder. During operation, there must be no danger from high voltage, either directly or due to the failure of a component. Cabinets should be interlocked so that high voltages are removed when the interior of the equipment is accessible. A repair man or maintenance man must be able to reach check points or remove or measure components without danger from high voltages or thermally hot elements. It is not enough to furnish labels warning of danger from high voltage.

Size and weight are not important in heavy industrial equipment, or even in console television sets, but they are most important in airborne equipment and some medical electronic devices, such as hearing aids. Compactness can be achieved to some extent by using miniature parts, but in general small size creates crowding, which causes other problems which must be taken into consideration. Crowding means parts are less accessible so that assembly and maintenance become more difficult. Also, with parts close together, heat dissipation may be a problem.

8-2 TYPES OF ASSEMBLIES

Even the simplest electronic equipment has a huge number of different components which must be arranged in some systematic order. The designer must lay out the components keeping in mind ease of assembly, human factors in operation, and accessibility for maintenance and repair. In addition, the equipment must meet all specifications. A logical solution to the layout problem is to arrange components in the order in which a signal travels through the circuits. A simple block diagram can be a guide. The electrical engineer will also assist here and may specify which lead-lengths must be kept short and which components cannot be close together.

If the equipment consists of a simple, small chassis on which all components are rigidly mounted, as in a portable transistor radio or a hearing aid, all assembly operations are usually done in *series*. Thus, each component is mounted in turn, and finally all are wired together. In more complicated equipment, it is possible to have individual subassemblies built simultaneously or in *parallel* and joined together later. Each subassembly may in fact be assembled serially, but the whole equipment is considered a parallel operation. An assembly procedure which can be composed of many parallel tasks is preferred over a straight series operation, since it permits more flexible use of workers.

The advantages and disadvantages of using subassemblies must be weighed for each job. As mentioned above, the obvious advantage is that subassemblies permit parallel assembly procedures. Subassemblies, especially plug-ins, simplify maintenance and repair. Space can be used more efficiently by using subassemblies, since there is access to areas which are inaccessible in a rigid single-chassis design. On the other hand, subassemblies require extra structural parts, more hardware, and some means of interconnection. The added cost of materials, however, may be more than offset by the savings in assembly time. Another factor, important in airborne applications, is the added weight of the extra hardware and structural supports.

It is possible to place two or more components into a single receptacle and treat this unit as a single component in production. Typically, three capacitors may be potted in one can, or a capacitor and resistor which are electrically connected may be encapsulated together. This permits some degree of parallel assembly in even the simplest circuits. A disadvantage is that when a single part in the receptacle fails, the whole receptacle must be be replaced, increasing the cost of repair.

Single chassis construction, as was pointed out above, requires piecemeal assembly, each part being permanently attached to the chassis in series. Even here, the designer should consider the possibility of using a *subchassis* in order to permit parallel assembly. A subchassis is simply an auxiliary surface to which components are attached. This subchassis is then attached to the main chassis later. Typically, if a chassis requires a partition, the partition may be made a subchassis. The subchassis may be fixed permanently or it may be removable to provide access to the interior.

Subassemblies may be put together in a variety of ways. The standard structure for large stationary equipment is the *relay rack* or *rack-and-panel* arrangement shown in Fig. 8–1. The open rack, shown in (a) consists of two vertical structural members firmly attached to a self-supporting base. Each individual chassis is mounted on a panel which is then attached to the rack by ordinary machine screws. In the most common size, the vertical uprights have holes spaced $18\frac{5}{16}$ in. horizontally and these holes are tapped with a #12-24 thread. The horizontal clearance between the uprights is $17\frac{1}{2}$ in. The panels are 19 in. wide with clearance holes which mate with the

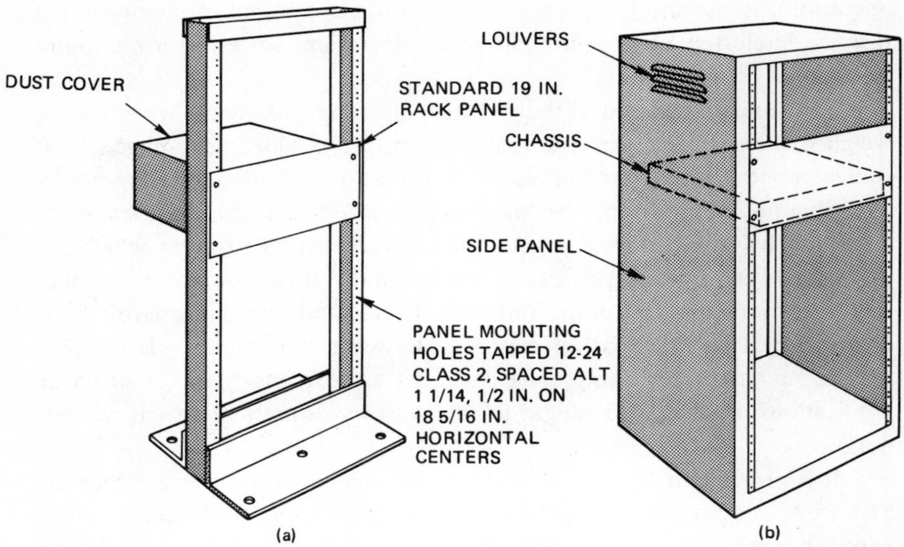

FIGURE 8-1 Standard Relay Racks

tapped holes on the uprights. This is called a 19-inch rack. A chassis with a maximum width of $17\frac{1}{2}$ in. is attached to each 19-inch panel. The vertical height of each panel varies from less than 3 in. to more than 20 in. The dimension chosen for the panel must exceed the vertical height of all components mounted on the chassis connected to it. A total of about 77 in. of panels can be mounted on a single rack. Shorter racks are available for smaller jobs.

The cabinet rack, shown in Fig. 8–1(b), is an enclosed cabinet with the front identical to the front of an open rack. When the panels are in place, the electronic equipment is entirely enclosed. The back of the cabinet may be a latched door for easy access to the interior.

Figure 8–2 is a photograph of a communications station built in rack-and-panel style. To improve the appearance, brackets have been added to the outside edges to hide the mounting screws. A typical chassis attached to a panel for mounting on a rack is shown in Fig. 8–3. This is a bottom view. The chassis is securely fastened to the mounting panel, shown at the bottom of the photograph.

Other assembly designs are shown in Fig. 8–4. The frame construction of (a) may be divided in any convenient fashion. Each subassembly fits into one space in the frame and may be mounted on its own subchassis or on a partition wall. The obvious disadvantage of frame construction is the added weight and cost of the structural members. The modular construction shown in (b) is advantageous when each subassembly can be packaged in its own enclosed box or module. Although the blocks are shown mounted on a chassis, this is not a requirement; they may be joined in any convenient fashion. Modular construction is particularly attractive when all modules are the same size so that stocking of different box sizes is eliminated. It is also advantageous when a basic system consists of one or more modules with additional capabilities furnished by adding other modules.

The hinged assembly in Fig. 8–4(c) permits easy access to the interior as well as permitting parallel assembly. Each hinged door contains one or more subassemblies in addition to the subassemblies in the main cabinet. The card file assembly, shown in (d), is used in printed-wiring construction, and is discussed in the next chapter.

The chassis shown in Fig. 8–5 is divided into separate compartments like the frame construction illustrated in Fig. 8–4a. The compartments on the right are for printed wiring cards. This is a photograph of the chassis for the receiver shown in Fig. 1–1 before wiring. It should be noted also that the front panel is a standard 19 in. panel for mounting on a rack as shown in Fig. 8–1.

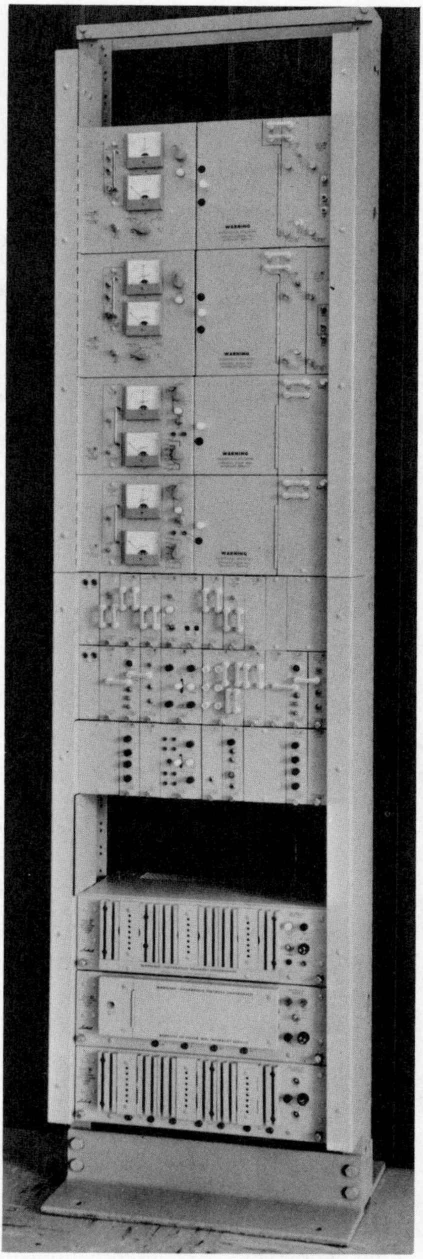

FIGURE 8-2 Rack-and-Panel Equipment (Courtesy of Lenkurt Electric Co., Inc.)

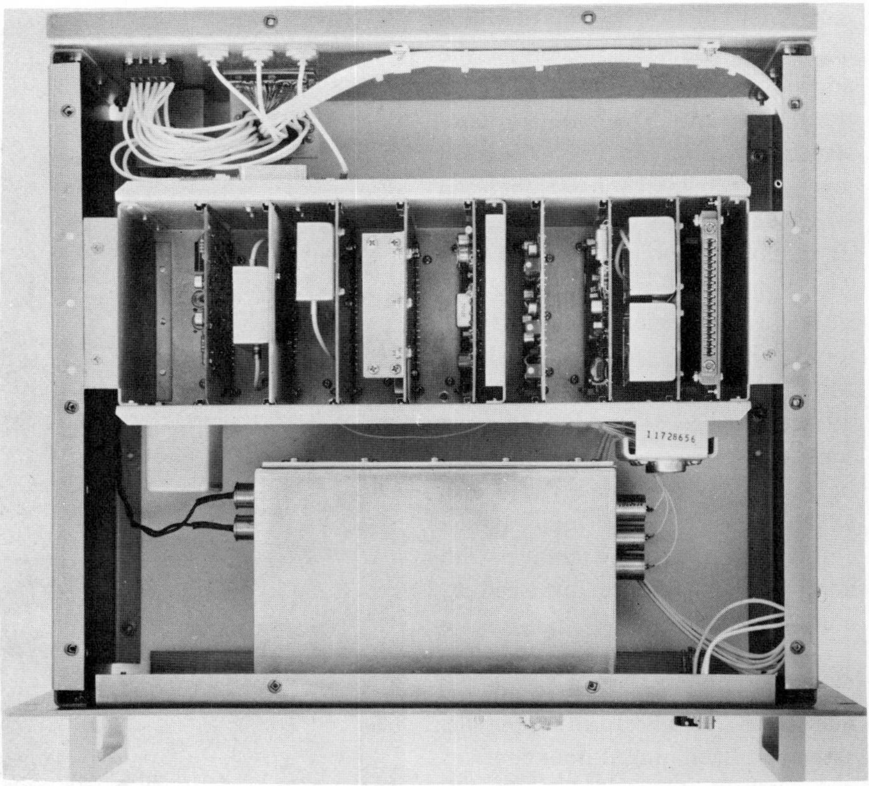

FIGURE 8-3 Panel with Chassis (Courtesy of CALIFORNIA MICROWAVE)

An example of hinged assembly is shown in Fig. 8–6. This is a photograph of a microwave communication terminal with subassemblies in hinged doors for ease in fabrication and maintenance. The upper box contains the transmitter, and the receiver is below. Figures 8–7 and 8–8 are photographs of the transmitter and receiver, respectively, with doors closed. A careful examination of Fig. 8–6 will show that this system, too, is mounted on a rack-and-panel assembly.

8-3 HARDWARE

Screws, nuts, washers, and lock washers are used to join parts in assemblies wherever a permanent bond is not wanted. Machine screws and bolts are available with many different head configurations in a wide variety of

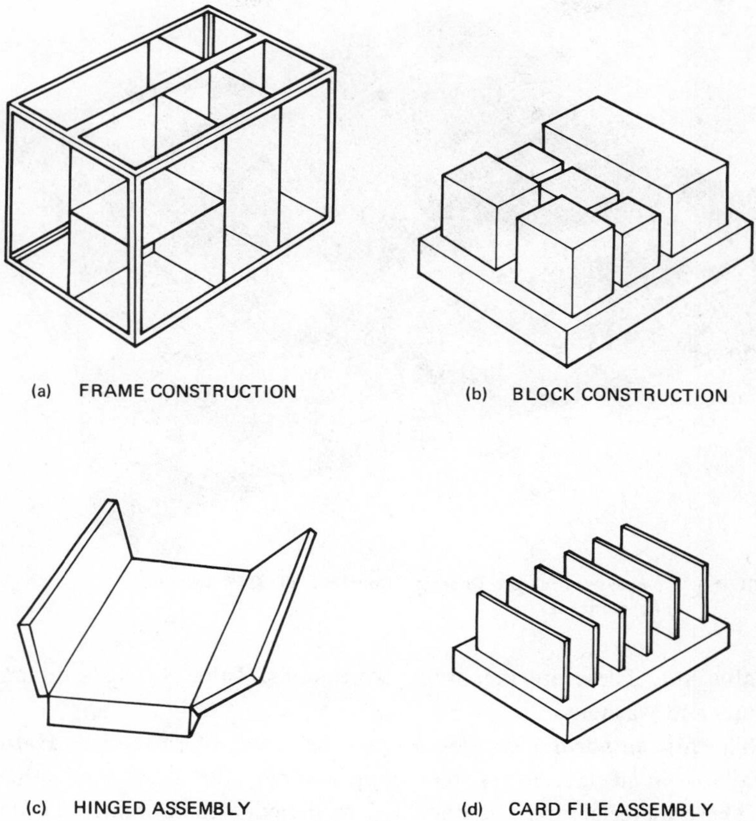

FIGURE 8-4 Typical Structural Designs

metals. Common head shapes are shown in Fig. 8–9. The head recess for any of the head shapes shown in (a) may be either the common screwdriver slot or the cross recess (Phillips head), both shown in (b). The set screw shown in (c) usually has a hexagonal or Allen-type recess. Although all head styles are used in electronic assemblies, the pan head is preferred and is generally selected unless there is a good reason not to use it. One reason for not using a pan head is a requirement for a flush mounting. Flat head screws are then selected to be installed in a countersunk hole. This requires an extra operation in fabrication; consequently flat head screws are not used unless flush mounting is a specific requirement. Usually an 82° flat head screw is used, but for thin materials 100° heads are also available.

Diameters and heights of heads for the different screw types are listed

FIGURE 8-5 Compartmentalized Chassis (Courtesy of GTE Sylvania Incorporated)

in Table 8–1. These are maximum dimensions. Table 8–2 lists dimensions for nuts and washers.

The first important consideration is selection of the metal. Hardware is available in steel, stainless steel, copper alloys, aluminum, and other metals. These materials may be uncoated or plated with cadmium, nickel, tin, and other materials. The important criterion is compatibility with the metal or metals being joined in order to avoid galvanic corrosion.

As was pointed out in Sec. 6.4, there is not much difference in strength between successive screw sizes. Thus, it is not important to stock all screw sizes, and most companies have a preferred or standard stock of selected sizes. The designer should use these sizes in the design. The minimum size screw which can be used depends on the weight of the part being supported and the number of screws holding it. Assuming a preferred stock of screws which includes #4, #6, #8, #10 and $\frac{1}{4}$ in., recommended sizes for different weights are shown in Table 8–3. For example, a 5-lb weight may be supported by four #8 screws or two #10 screws.

Screws should protrude at least $1\frac{1}{2}$ threads beyond the nut and up to $\frac{1}{8}$ in. more than this. In tight spaces, they may be shorter to avoid interference. In blind-tapped holes, the screw should engage about ten threads.

FIGURE 8-6 Hinges Assembly (Courtesy of Farinon Electric)

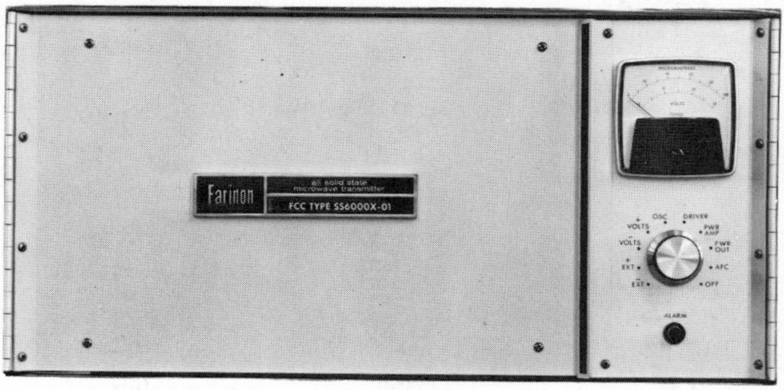

FIGURE 8-7 Transmitter (Courtesy of Farinon Electric)

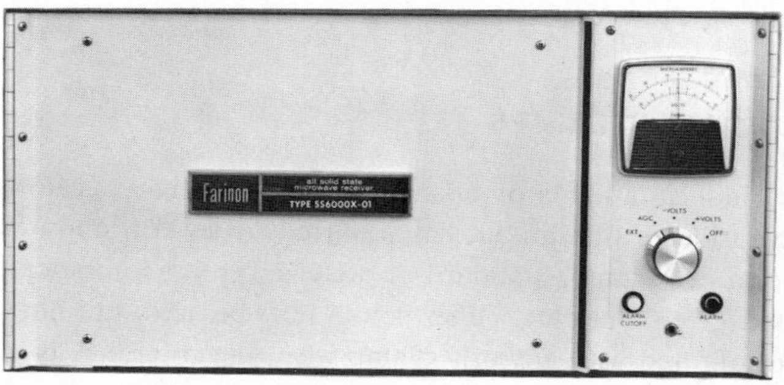

FIGURE 8-8 Receiver (Courtesy of Farinon Electric)

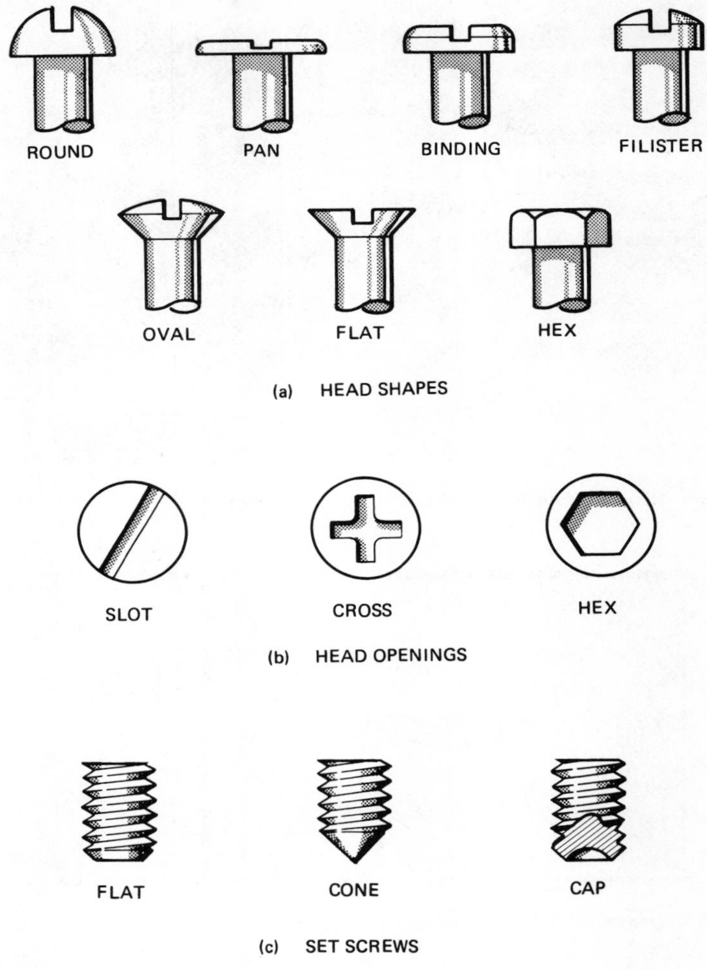

FIGURE 8-9 Screw Fasteners

8-4 INDICATOR LAMPS

Jewel indicator lights or neon bulbs are used to indicate that circuits are in operation or that a fault has occurred. A jewel window may be used in front of a neon lamp. Standard or preferred colors for indicator lights are red, amber, green and blue. Other colors may be used, but not white, since it may be confused with any other lamp when the color disk is inadvertently omitted.

Selection of colors for indicator lights may be arbitrary or may be standard in a particular company. In general, red indicates danger. It may be used to indicate that high voltage has been applied. Green indicates normal operation. Amber may be used for intermediate steps in the operation such as indicating that voltage is applied to filaments or blowers.

8-5 WIRING

After the components have been mounted in their proper places, they are interconnected by wiring. There are two general methods of doing this. The first is *point-to-point* wiring, which, as its name suggests, consists of connecting individual wires between components or between a component and the chassis. The second method is *harness* wiring in which all wires are cut to length and laid out on a wiring board. They are then tied together to form a harness which is then laid into the assembly, and the wire ends are fastened to the proper terminals.

In both cases, the wires are finally soldered to the proper junction point. Soldering may be done manually, using an iron or torch, by dip brazing, or by induction, depending on the complexity of the assembly. If a hand iron is to be used, the junction points must not be too heavy or else heat will be conducted away and make soldering difficult. Point-to-point wiring may be simpler for small operations, but harness wiring permits parallel assembly in that the harness can be made at the same time parts are being mounted or even earlier.

When many wires are to be connected to a single terminal point, heat from the iron may melt the solder and loosen wires already connected as new wires are attached. In general, the number of wires attached to any terminal should be held to a minimum and in no case should exceed three.

Regardless of the type of wiring to be used, the designer should prepare a *wiring diagram* and a *wire list*. The wiring diagram for point-to-point wiring simply shows each wire as a solid line connecting two points. The wires are identified by number and, in some cases, by color also. On the wire list, each length of wire is listed separately, along with complete descriptive information. This includes the wire size, the length, the color of the insulation, and the two points that it connects. A typical form for a wire list is shown in Fig. 8–10.

For harness wiring, the designer should prepare a scale model of the harness. This may be used as a template to fasten to a wiring board to simplify manufacturing the harness. Ideally the harness should be complete in itself with connectors fastened in place before insertion in the chassis. The assembly should be designed so that the harness can be laid in place and the connectors attached to the chassis without having to pass the harness through small holes.

Changes do occur in systems as new techniques are developed or improved components are invented. To anticipate changes without requiring complicated redesign, it is desirable to use connectors having more pins than required for the wiring. This also simplifies repair.

Wires should be parallel whenever possible in order to present a neat appearance. They should be isolated from moving parts and components that radiate heat. When a wire passes through a small hole in a metal wall, it may be chafed from vibration due to motion or handling. Suitable protection in the form of bushings or grommets should be provided.

In order to identify the two ends of a wire when there are many wires

								DRAWING NUMBER		
	WIRE					FROM		TO		
SYMBOL	TYPE	LENGTH IN.	AWG	COLOR	REF. NO.	LOCATION	NOTES	LOCATION	NOTES	CIRCUIT FUNCTION

FIGURE 8-10 Wire List

in a group, colored insulation is used, and every wire in a group must be colored differently. Color coding of wires is mainly for differentiation, although some companies use colors to indicate the functions of the wires. Thus, wires connected to ground might be black, positive high voltage red, anodes blue, etc. With the complexity of modern electronic equipment, a functional color code requires more trouble and time than it is worth.

Since there are only a few different colors, and since the number of wires in a group may exceed the numbers of colors available, insulation is made with tracers of contrasting colors in otherwise solid-colored coverings. To increase the number of available coverings, insulations may be of one solid color, a solid color with one contrasting tracer, or a solid color with two tracers. In order to identify the color scheme, a numerical color code is used. This is based on the resistor color code and is described in Table 8–4. Thus, a red wire has a code designated 2; a white wire with a green tracer would be designated 95; and a white wire with yellow and blue tracers would be 946, but not 964. When two tracers are used the third digit is always higher than the second. Some companies use a three-digit color code for all wires. In this case a red wire becomes 222. This should indicate a red covering with two red tracers, but since there is no contrast between the tracers and the main color, it indicates a solid color. By the same reasoning, a white wire with a single green tracer is designated 959. On wiring diagrams the numerical color code is usually used in preference to calling out the colors by name.

Wire size is designated by a gauge number which specifies the diameter of the wire. The standard is the American Wire Gauge (AWG), which is also called Brown and Sharpe (B & S) wire gauge. A listing of sizes by gauge number and other pertinent information is given in Table 8–5. Sizes of wire with insulation is given in Table 8–6. In the American Wire Gauge, the circular area decreases by a factor of two for an increase of three wire sizes. Circular area is given in *circular mils,* which is simply the square of 1000 times the diameter in inches. In selecting a wire size for specific wires in the circuit, the designer must consider current carrying capacity and ease of handling the wire. Thus, even if negligible current is involved, the designer would not specify #46 wire, since it is too fine to be handled easily.

Heavy solid wire is difficult to bend and may be damaged by vibration due to its rigidity. It is possible to twist several strands of finer wire together to get an equivalent cross-sectional area but with more flexibility. This is called *stranded wire.* For example, #16 has a cross-sectional area of 2583 circular mils. Common stranded wire has seven or ten strands. Thus, seven strands of #24 wire, which has a cross-sectional area of 404 circular mils, will have a total of 2828 circular mils and will be able to carry whatever current #16 wire can. Also, ten strands of #26 wire will have a total cross-sectional area of 2541 circular mils and this can be used in place of #16 wire. It is slightly less in area but within the engineering safety factor. The two stranded wires would be designated 7/24 and 10/26, repectively. Some common stranded sizes which are used to replace solid wire are given in Table 8–7.

In electronic equipment the most common wire sizes range from #10 to #28. Selection is usually based on current carrying capacity. Recommended sizes for different currents are given in Table 8–8. These values are not critical and contain a reasonable margin of safety. Usually stranded equivalents of the designated wire are selected.

9

PRINTED WIRING BOARDS

The complexity of modern electronic equipment with its myriad of components and interconnections has a serious effect on reliability. Workers do get tired and make errors, and the sheer numbers of interconnections required make error-free assembly a rarity. Printed wiring eliminates much of the manual assembly and thus increases reliability. An incidental benefit is a reduction in size and weight.

In some complicated systems, more than 50 per cent of the cost of manufacture is spent for production of *printed wiring boards*. This segment of the electronics industries has grown to such an extent that there are now many companies whose sole business is supplying printed wiring boards for other manufacturers. For simplicity, these boards are referred to as PWB.

Printed wiring is sometimes erroneously called *printed circuitry*. In printed wiring only the conductors are *printed*, and components are attached to the printed wiring. In printed circuits, components as well as connectors are printed. *Hybrid circuits* are combinations in which some components are printed and others are added later.

9-1 ADVANTAGES AND DISADVANTAGES OF PWB

As mentioned above, the first advantage of printed wiring over point-to-point wiring is increased reliability. Once a design is fixed and a correct photographic film made, the conductors are exactly reproducible without possibility of error. It is still necessary to attach components such as capacitors, resistors, and transistors to the board. This is known as *stuffing* the board and is usually accomplished manually, although automation is moving into this area also.

Reduction of weight and size can be dramatic. In some cases weight has been reduced to one-tenth of what it would be with conventional mounting and point-to-point wiring. Along with weight reduction there is a corresponding cost savings because of automated or semi-automated production. The number of interfaces between worker and work is reduced.

Accessibility of all parts is improved. This means that inspection is

FIGURE 9-1 Combining Case (Courtesy of Hewlett-Packard)

more easily accomplished, trouble-shooting is simplified, and all portions of the circuit are easily reached for servicing. Figure 9–1 is a photograph of an assembly containing twenty data amplifiers. Each board, containing two amplifiers, is removable from the case for easy access to the circuit. This method of assembly was illustrated in Fig. 8–4(d).

Not the least of the PWB advantages is the improved appearance of printed wiring over point-to-point wiring. This is important, since the vast majority of those involved with a piece of electronic equipment, such as purchasing agents, supervisors, and the like, do not use the equipment but only inspect it visually.

PWBs do not solve all problems, and they do have some disadvantages. Heat dissipation is poor, and as a consequence thermal design may be complicated. In some cases where adequate heat dissipation from part of the circuit is too difficult to obtain in a printed wiring board, that part may be constructed on a separate subchassis with point-to-point wiring, while the rest of the circuit utilizes PWBs.

Another disadvantage is rigidity of the design. Once the design is set, it cannot be revised except by discarding the complete board and substituting a different one. The rigidity of design also forces some electrical and mechanical compromises. Heavy components cannot be mounted except with great difficulty.

Although parts are readily accessible for trouble-shooting and most repairs are not too difficult, an intermittent fault caused by peeling and movement of a copper conductor on the board is very difficult to find and impossible to repair. In a well-made board the copper should not peel from the base material, but occasionally this does happen because of poor bonding originally.

9-2 MATERIALS

Printed wiring boards are made of a nonconducting laminated material to which a thin sheet of copper has been bonded on one or both sides. The laminate consists of a resin or plastic with a filler of paper, fiber, or glass. Although paper-filled laminates are most common in electronic equipment aimed at the consumer market, they do not meet military specifications and thus are not used in military electronic equipment.

Laminates have been standardized and are described in the specifications of MIL-P-13949 and of the National Electrical Manufacturers Association (NEMA). Table 9–1 lists some of standard laminates with both their NEMA and military designations, where applicable. Materials which are flame resistant have a NEMA designation beginning with FR. The materials come in an assortment of colors and vary from translucent to opaque. Color is not a reliable guide to identification. The paper-filled materials are by far the cheapest, and the glass-filled polytetrafluoroethylenes are the most expensive.

Physical properties which should be considered when selecting a material for PWBs are:

1. PEEL STRENGTH. This is a measure of the bond between the copper and the laminate. Good peel strength indicates that the PWB will withstand abuse without having the copper conductors separate from the base material. Good bonding also aids in heat dissipation.
2. LEAKAGE RESISTANCE. This is a measure of the electrical resistance between adjacent conductors on the PWB. Since leakage resistance tends to decrease as temperature and humidity increase, tests should be performed for this parameter under the worst environmental conditions which will be encountered.
3. FLEXURAL STRENGTH. This is a measure of the mechanical resistance of a PWB to bending and vibration. A weaker material may still be used but may require more support to prevent damage.
4. WATER ABSORPTION. This is a factor in increasing the leakage resistance because of increased humidity. It is also an indication of the tendency of the material to absorb chemicals during etching and plating processes.
5. FLAMMABILITY. Flame-retardant materials are available, and these will not burn even when a flame is applied directly to them.
6. DIELECTRIC CONSTANT. This is an electrical parameter which affects the choice of size of copper areas on the PWB.
7. DISSIPATION FACTOR. This is a measure of power lost in the material.

The base material is available in standard sizes which are multiples of $\frac{1}{32}$ in. A practical size for most applications is $\frac{1}{16}$ or 0.062 in. Tolerance on the thickness is specified in MIL-P-13949 and is shown for some thicknesses in Table 9–2. It should be noted that the thickness specified in the table is the nominal overall thickness, but the tolerances are loose enough so that these values could also be taken as the thickness of the base material alone. In general, epoxy-glass laminates are better than any of the paper-filled materials, and polytetrafluoroethylenes are best of all.

Of special importance is peel strength, and paperfilled materials are generally poorest in this regard. Even though they are cheaper, it may be false economy to use such a material if a card is spoiled because the copper separated from the base material. In order to avoid having to stock many different materials, a company will usually standardize on one preferred material and one preferred size. For inexpensive consumer products such as radios, this might be a paper-filled material such as XXXP. However, if any work is being done for the government or if the equipment has to be more reliable, a good choice is FR-4 (GF) or FR-5 (GH). These materials rate good to excellent in all physical properties listed above and are reasonably priced. A thickness of 0.062 in. is usually preferred.

9-3 CLADDING

The board material is supplied with copper bonded to one or both sides. The thickness of the copper is not specified directly, but instead the weight of a square foot of copper at that thickness is given. Thus, the cladding may be indicated as 0.5-oz copper, 1-oz copper, etc. The relationship between weight and actual thickness is given in Table 9–3.

For most applications, 2-oz copper is satisfactory. The heavier the copper, the higher is the peel strength, and the more reliable is the PWB. However, heavy copper cannot be etched as finely as lighter grades. Thus, for fine detail 1-oz copper or even 0.5-oz may be used, but 2-oz is used generally. If high temperatures may be a problem, heavier coatings may be used as heat sinks.

When specifying a material to meet MIL-P-13949, the designer calls out the material in an alphanumeric code. First, the letters FL indicate that the material is a foil-clad laminate. This is followed by a hyphen and two letters designating the MIL material, as indicated in Table 9–1. Then three digits indicate the thickness in thousandths of an inch, the letter C signifies copper, and finally the weight of copper is shown for each side of the board. For example, Type FL-GH062C2/2 would mean the material is GH, 0.062 in. thick, with 2-oz copper on both sides. If cladding is wanted on one side only, one final digit is used after C instead of two. It is also possible to obtain material with different weights of copper on each side.

9-4 FABRICATION OF THE BOARD

In most customary machine processes, the drawing need not be an accurate scale representation of the finished part. In fact, the machinist is cautioned to follow the dimensions given and not try to scale the drawing.

The printed wiring board is different in that the drawing itself is used to make a template for the fabrication of the board, and the PWB can be no more accurate than the drawing. In order to have a better appreciation of the design problems, the designer should be aware of the total process of fabrication of PWBs.

A printed wiring board consists of a thin laminated material, as described in Sec. 9–2, on which conductors of thin copper foil are bonded. The conductors are not actually printed on the board. In practice, the laminate is furnished with a thin copper foil completely covering the surface. This copper sheet is etched away except where the conductors are required. The desired pattern is *printed* on the copper in a manner similar to offset printing, and thus the boards are called printed wiring. The complete fabrication process consists of etching the pattern on the board, drilling holes for mounting components, and mounting and soldering the components in place.

The basic etching process is illustrated in Fig. 9–2. A photograph is made of the designer's drawing and that photograph is printed on the copper foil in an organic ink called a *resist*. The resist is acid-resistant. As shown in (a), the resist is in the form of the desired pattern. This is *direct* or *positive* printing. The board is then placed in an acid bath where the copper which is not protected by the resist is etched away, as shown in (b). The resist is then dissolved leaving the finished pattern, as shown in (c). Holes are now drilled or punched as shown in (d). For ease in soldering components to the conductors later, the board at this point may be solder-wiped in a hot tin-lead bath.

The tin-lead coating may be electroplated onto the copper instead of being wiped on. This is preferable and necessitates a different fabrication procedure, illustrated in Fig. 9–3. The resist in this case is applied to the

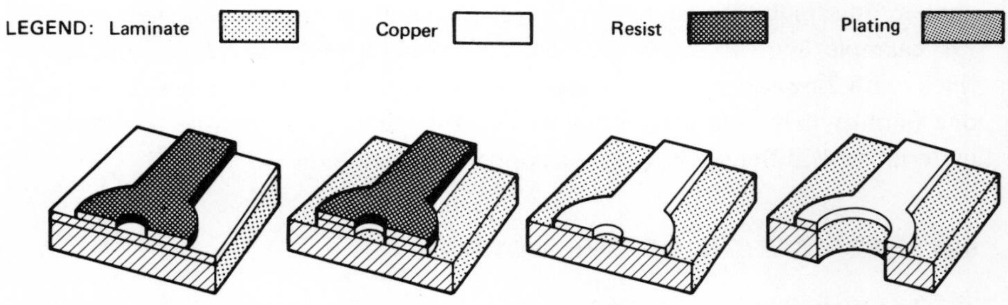

(a) RESIST APPLIES (b) ETCHED (c) RESIST REMOVED (d) HOLES PUNCHED

FIGURE 9-2 Basic Process

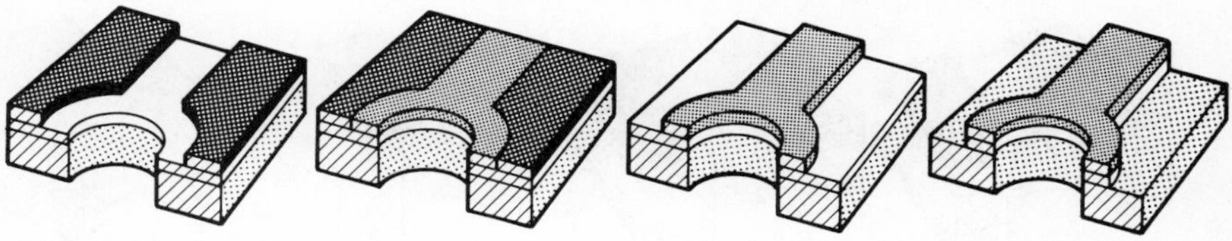

(a) REVERSE PRINTED (b) PLATED (c) RESIST REMOVED (d) ETCHED

FIGURE 9-3 Plated and Etched

parts of the copper sheet *which are to be removed*, as shown in (a). This is *reverse* or *negative* printing. Now the board is plated, and here the resist acts to keep the plating solution off unwanted areas. The result is shown in (b), and again with the resist dissolved away, in (c). The board is now placed in an acid bath which etches away unwanted copper, leaving the plated conductors as in (d). The plating on the conductors acts as a resist in the etching process.

In Fig. 9–2, holes were indicated as being punched after the etching process, and in Fig. 9–3, the holes were shown punched before or after the printing and etching. However, to avoid scratching the plating during punching it is preferable to have the holes done before the boards are plated, as shown in Fig. 9–3. The holes in boards treated as shown in Fig. 9–2 may be drilled or punched at any time before the hot-solder wipe.

The processes shown in Figs. 9–2 and 9–3 may be used when conductors are placed on only one side of the board. This is the so-called *single-sided board.* It is also possible to have conductors on both sides of the board, starting with a material which has copper bonded to both sides. The *double-sided board* may be dictated by economy or a problem of insulating conductors from one another. On the single-sided board, components are mounted on the opposite side from the conductors. On the double-sided board, all components are mounted on one side, although conductors are on both sides.

The process for double-sided boards is shown in Fig. 9–4 and is similar to the plating process shown in Fig. 9–3, but the walls of the holes are plated also. The process begins with the holes being punched or drilled, and the resist is reverse printed, as shown in Fig. 9–3(a). The holes are now sensitized so that when the copper is plated the holes are plated also. This is shown in Fig. 9–4(b). The resist is removed (c), and the unwanted copper is etched away (d). The plating acts as a resist both on the copper surfaces and in the holes.

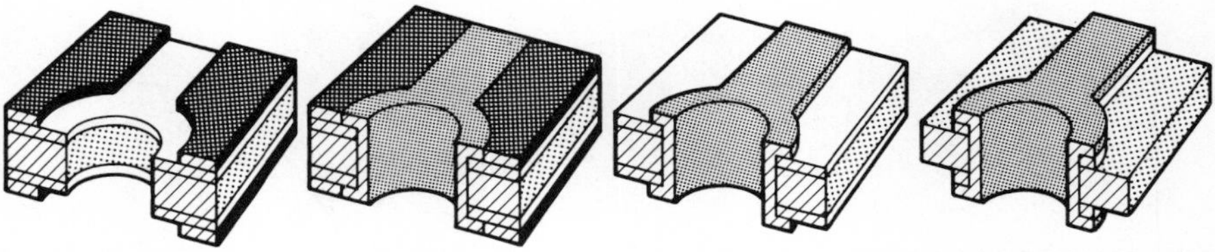

FIGURE 9-4 Plated-Through Holes

9-5 ARTWORK

The designer must plan how components are to be mounted on the PWB and how they are interconnected. Thus, he must lay out the mounting holes and the conductor paths and indicate how components are to be mounted. Layouts of PWBs are quite different from other electronic packages in that everything must be planned in essentially a single plane.

The size of the boards and their positions in the equipment are usually determined by mechanical requirements. Boards up to $\frac{1}{16}$ in. thick should be supported every four inches. For practical considerations this means that a $\frac{1}{16}$ in. board should not exceed four inches in length.

The electrical design has been selected by the electronics engineer on the project, and he has furnished a schematic diagram and some pertinent information about voltages, currents, and leads which must be short or isolated. The designer makes a rough layout of the board, indicating the positions of all components and how they are connected. This is done in pencil to permit changes and is usually at least twice size. Conductors are indicated by dotted lines when they are on the side opposite to that on which components are mounted, and by solid lines when they are on the same side, which would occur on double-sided boards.

The locations on the drawing should be based on a modular grid. If MIL-275B applies, the preferred unit of length on the grid is 0.100 in., but 0.050 in. and 0.025 in. are also permitted. This is good practice even in non-military equipment. All components are arranged so that their mounting holes are located at grid intersections. If a component must be mounted so both holes cannot be at grid intersections, then one hole must be so located, and the other must be dimensioned in x and y coordinates from the first.

The finished pencil layout is an oversize scale model of the PWB. It is placed on a light-table and covered with a stable, transparent material. Black tape is attached to the transparent sheet wherever a conductor is indicated, and black discs or other shapes are attached as needed. Tape is available in a number of standard widths and also in other precut standard shapes, such as circles or ovals. The preferred width of tape is $\frac{1}{16}$ in. on a twice size drawing, which reduces to a conductor width of $\frac{1}{32}$ in. on the finished board. Narrower tape may be used if the board is crowded. The minimum spacing between adjacent conductors is equal to the width of the conductors. A large photograph of the *tape-up*, called a *master drawing*, is made for future use.

Films used in the fabrication process are the same size as the finished PWB. These may be photo-reduced from the tape-up or from the master-drawing. Usually the tape-up is discarded after the first life-size films are made so that subsequent films are made from the master drawing. These films may be positives or negatives, depending on the type of board and the process. For a single-sided board which will have a hot-solder wipe, a negative is needed. If the board is to be plated instead, a positive is used. One film may be used for several boards of the same type.

In addition to the master drawing, the designer must prepare a *schematic diagram*, a *printed wiring board drawing* and an *assembly drawing* for each board. The schematic diagram is a finished copy of the rough schematic supplied by the electronic engineer. The PWB drawing shows the size and shape of the board, calls out the board material, and indicates the sizes and locations of all holes. The assembly drawing shows how components are mounted on the PWB. All components are also listed in a separate parts list.

9-6 CONDUCTORS

In laying out the conductors on the board, the designer strives to route them in the shortest paths possible consistent with the electrical design. Nevertheless, some thought should be given to "eye-appeal," and wherever possible a neat, symmetrical arrangement of conductors is preferred. Sharp internal angles should be avoided, since the acid in the etching solution may not penetrate to the desired point. Sharp external points should also be avoided since copper tends to peel at a corner. Thus, interior angles should have a fillet, and exterior corners should be rounded. Interior angles less than 90°, even with a fillet, should be avoided. Figure 9–5 shows examples of good and bad conductor design illustrating these points. The radius at both inside and outside corners should be at least equal to the width of the strip.

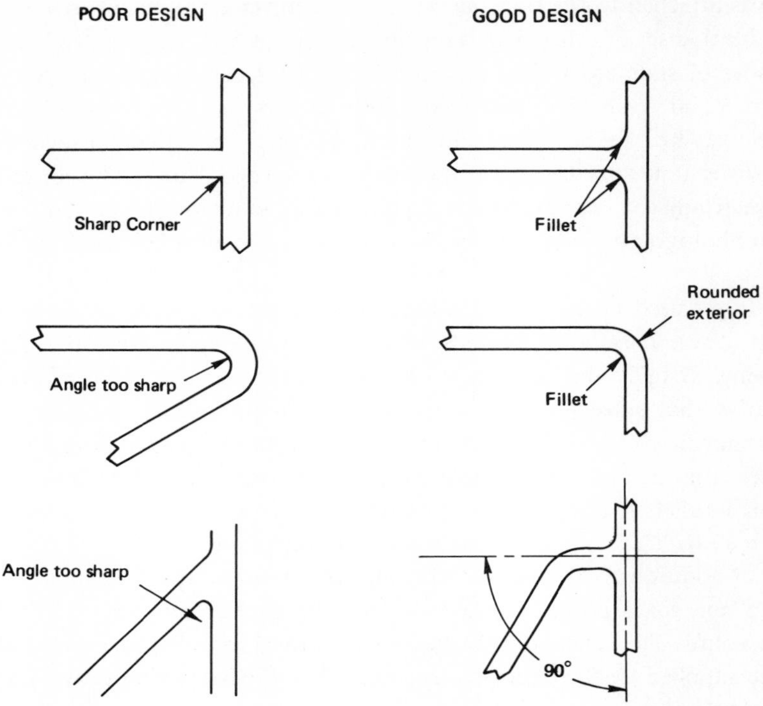

FIGURE 9-5 Conductor Angles

A *terminal pad* must be provided wherever a component is to be attached to a conductor. A pad is simply an area of copper, usually round, larger than the width of the conductor. The mounting hole is drilled or punched through the approximate center of the pad, leaving an annular ring to which the component lead is soldered. Pads are also required at points where conductors on opposite sides of the board are joined. If conductors must pass between two pads or other large areas, without making contact, they should maintain maximum and equal spacing from both. Examples of good and bad routing are shown in Fig. 9–6.

When large copper areas contain one or more pads for soldering a lead to a conductor, during the soldering operation, the hottest part of the area will be in the center, and the solder will tend to flow there. If leads are to be soldered to points off-center, the conductive area should be shaped to equalize the temperature at the solder points. In general, if the pads joined by a conductor are less than a pad diameter apart, the width of the conductor may be the same as the pad diameter, but for greater distances it should be reduced.

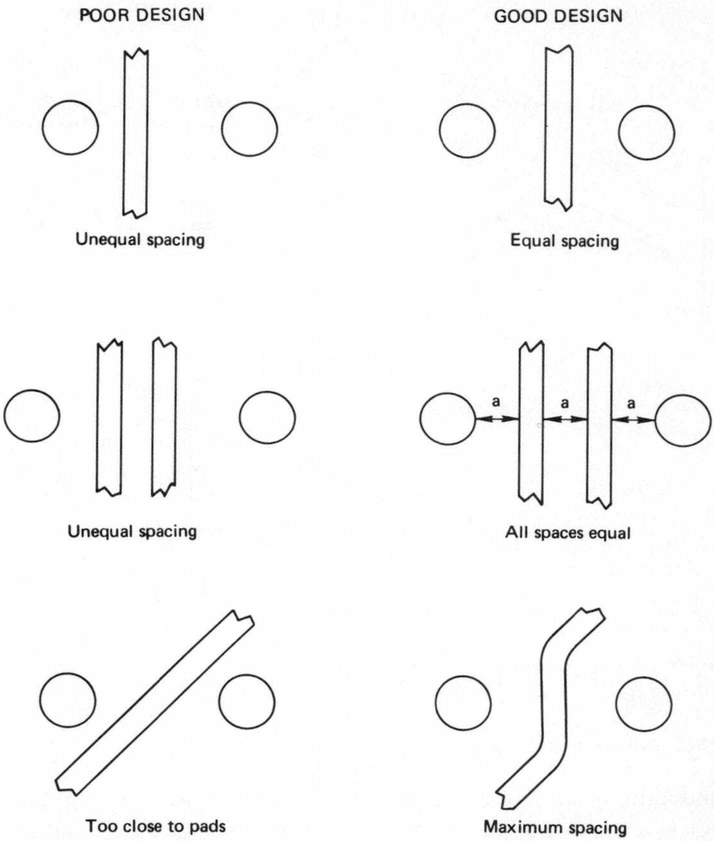

Figure 9-6 Conductor Routing

If three or more pads are spaced on a single conductor, they should be at least two diameters apart, or heat will concentrate at the center pad. If pads of unequal size are connected, heat will run to the larger pad unless they are spaced sufficiently. Acceptable designs and minimum spacings are shown in Fig. 9–7. In the figure, D is the pad diameter. All interior corners between conductors and pads should have radii.

The current-carrying capacity of conductors is a function of their size. *Heat* is the limiting factor. A conductor which is thin and narrow has a relatively larger resistance and therefore will produce more heat due to the I^2R loss when current flows.

The *temperature rise* for a given current and conductor size can be calculated from the graphs of Fig. 9–8, which is taken from MIL-STD-275.

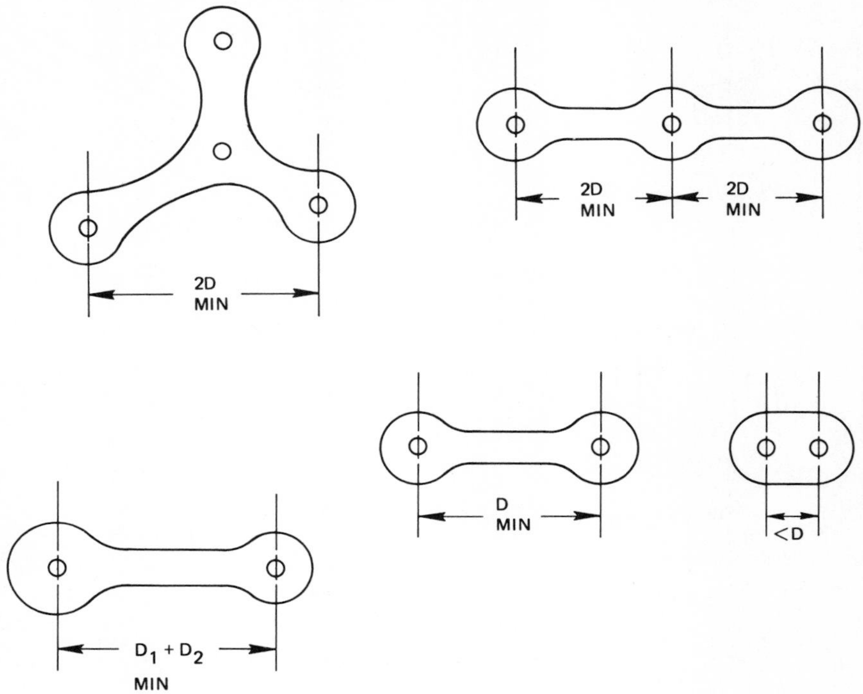

FIGURE 9-7 Pad Spacing

The conductor width is entered on the ordinate of the lower graph. Where it intersects a curve for the weight of copper, the cross-sectional area may be read on the abscissa. Using this area and the current, the *temperature rise* may be read on one of the curves of the upper graph. These values include a 10 per cent derating factor to allow for variations in material. The *operating temperature* is the sum of the temperature rise and the ambient temperature. This must not exceed the maximum rated operating temperature of the material, which is about 220°F for paper-filled epoxy, 250°F for glass-filled epoxy, and 400°F for polyterafluoroethylene materials.

When several parallel conductors are close together, the total temperature rise will be the sum of those caused by currents in all the conductors. Figure 9–8 may be used to calculate the total rise by using the sum of all the currents as the current value and the sum of all the cross sections as the cross-section value.

The minimum spacing between conductors is equal to the conductor width. A good, practical width for most aplications is 0.032 in., although

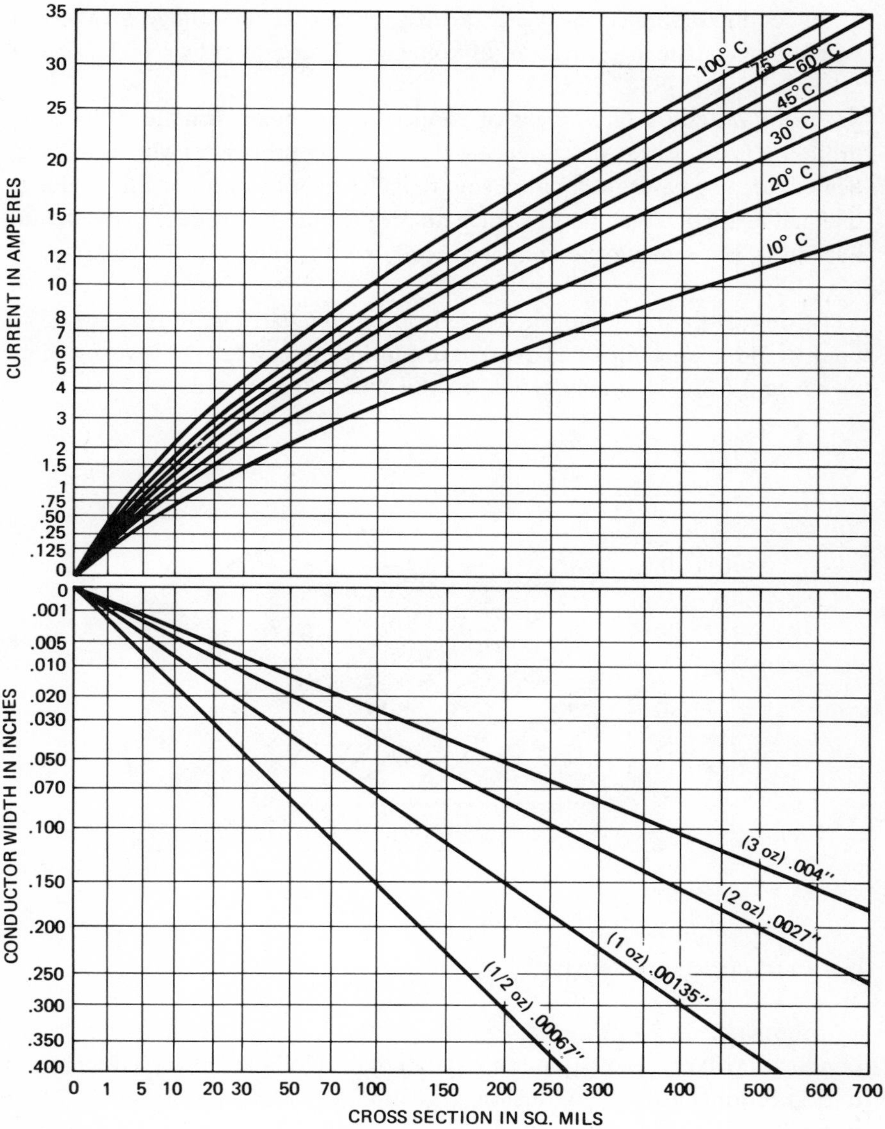

FIGURE 9-8 Conductor Dimensions

this is not necessarily the minimum permissible width. In dealing with the Federal Government, contracts sometimes specify minimum width.

The minimum spacing between conductors is also affected by the voltage difference between adjacent conductors. When voltage differences are high, conductors must be spaced farther apart to avoid voltage breakdown. Minimum spacing as a function of voltage difference is shown in Table 9–4, taken from MIL-STD-275.

Land areas are large areas of copper on the board usually intended as ground planes or heat sinks. During the soldering process or when otherwise heated, areas greater than half a square inch are apt to peel or blister due to unequal heating. To avoid this, such an area should be broken up by etching lines in it, but leaving the area electrically continuous. This is illustrated in Fig. 9–9.

Conductors should not be located closer than 0.010 in. to the edge of the board, and a spacing of at least 0.030 in. is preferable. However, ground planes and heat sinks may extend to the edge of the board.

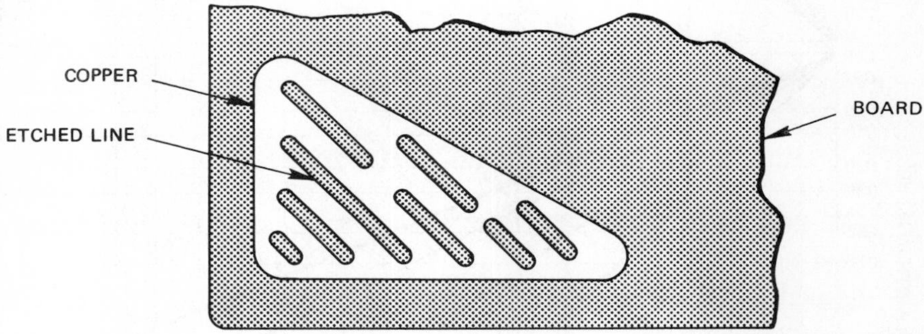

FIGURE 9-9 Land Area

9-7 HOLES AND PADS

Mounting holes are used to connect the leads of components to the conductors. Every lead requires a separate hole. Holes are also used to permit a connection between conductors on opposite sides of the board. These holes may have a conductive plating or a metal eyelet connecting the two surfaces. If there is no conductive path, the hole is said to be *unsupported*. Usually, unsupported holes are used only on single-sided boards. The diameter of an unsupported hole must not exceed the diameter of the lead to be inserted by more than 0.028 in. For supported holes, the hole diameter or inside eyelet diameter can be as much as 0.035 in. greater than the diameter of the lead. These are military specifications which are easy to meet, although in non-military applications, hole diameters up to 0.050 in. greater than lead diameters have been satisfactory.

The number of different hole sizes on a board should be kept to a minimum. Thus, if different lead diameters are used, some holes may be only a few thousandths greater than their lead diameters, while others go all the way to 0.035 in., thus permitting one hole size to be used for many different lead diameters. If the board is to be drilled or punched on a numerical machine, the number of hole sizes must not exceed the number of different tool sizes available. In planning hole sizes for a numerically controlled machine, the designer must remember that the reference holes on the card are made with one of the available tools.

A terminal pad must be used wherever a conductor intersects a hole. The pad has two purposes. It prevents the copper from peeling when the hole is drilled and it is used as a terminal to which a lead or eyelet can be soldered. The minimum diameter of the terminal pad depends on the application, but there is no upper limit. For unsupported holes, the pad must be at least 0.040 in. greater than the hole diameter. For plated-through holes, the pad must be at least 0.020 in. greater than the hole diameter. If an eyelet or solder terminal is inserted in the hole, the hole diameter must be *no more than* 0.010 in. greater than the outside diameter of the insert, and the pad must be at least 0.020 in. greater than the diameter of the flange on the insert.

When it is impossible to lay out the conductors on a one-sided board without an unwanted intersection, a jumper bridge may be used. This is shown in Fig. 9–10. A piece of wire is mounted on the component side of the board so that it connects the two parts of one conductor while bridging one or more other conductors.

Interfacial connections between conductors on opposite sides of the boards are made through holes. Although a plated-through hole does have a continuous electrical path from one side of the board to the other, it is not a reliable connection. Approved methods of connecting opposite sides of the board are jumper wires, metal eyelets, solder plugs, and stand-off terminals.

A *jumper wire* is a piece of uninsulated solid wire which passes through the hole and is soldered to the pads on both sides. The hole itself may be unsupported or plated-through. Figure 9–11 shows how the wire is bent or clinched in place and soldered. The direction of clinch is optional. Since the wire must be soldered to both sides of the board, the component side of it must be soldered by hand.

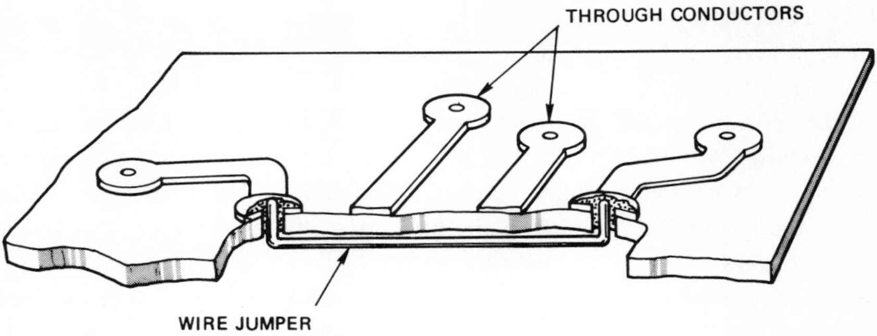

FIGURE 9-10 Jumper Bridge

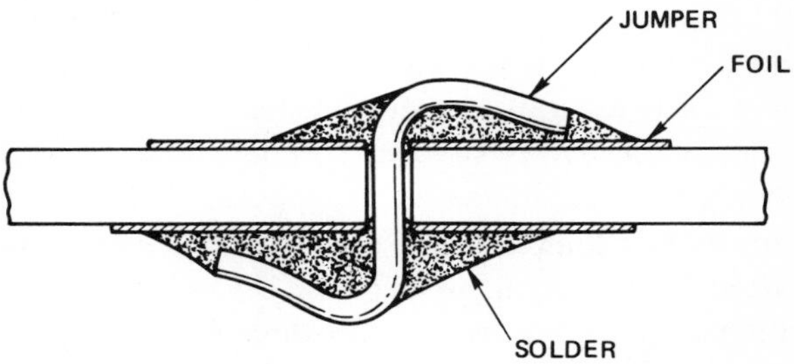

FIGURE 9-11 Jumper Wire

Eyelets are sometimes used instead of plating through the holes in a board. Approved plated copper eyelets have funnel flanges on each end. Unlike plating, the eyelet does furnish a satisfactory interfacial connection. However, if a single hole is plated through, they all are since the additional cost of sensitizing all the holes is negligible. If eyelets are used instead, they should be soldered on both sides in order to ensure that they do not loosen from thermal effects or mechanical vibration. Figure 9–12 shows an interfacial connection using an eyelet.

When a hole is used only as an interfacial connection, and no component lead or other wire is to be connected at that point, a reliable connection can be made simply by filling the hole with a plug of solder. This type of interfacial connection is shown in Fig. 9–13. The hole must be supported by plating through it.

A *stand-off solder terminal* may be mounted in a hole and soldered to conductors on both sides to make an interfacial connection. This is illustrated in Fig. 9–14. A stand-off terminal is used on both single and double-sided

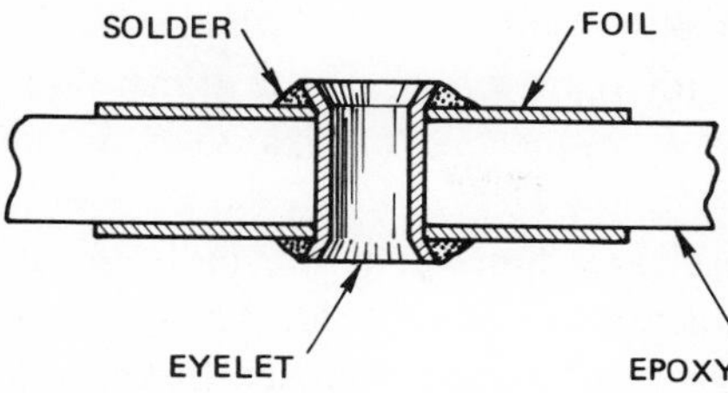

FIGURE 9-12 Eyelet

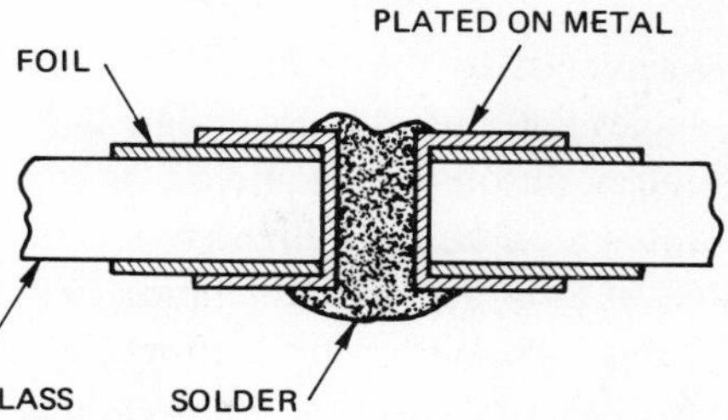

FIGURE 9-13 Solder Plug

boards whenever a wire or component lead may be subject to frequent soldering and unsoldering during maintenance and testing. As many as three wires may be soldered to a single terminal. Since the terminal is on the component side of the board, leads connected to it must be soldered manually.

When holes are *punched* in paper-filled materials such as XXXP or FR-2, the resiliency of the material causes them to spring back after the punch is removed. Thus, the hole is smaller than the punch. The amount of springback increases with increasing thickness and is about 0.001 in. for every sixty-fourth of an inch in thickness up to a thickness of $\frac{1}{16}$ in. Above this value there is an additional 0.002 in. for each sixty-fourth of an inch. The punch sizes called out on the drawing should thus be larger than the finished holes by these amounts.

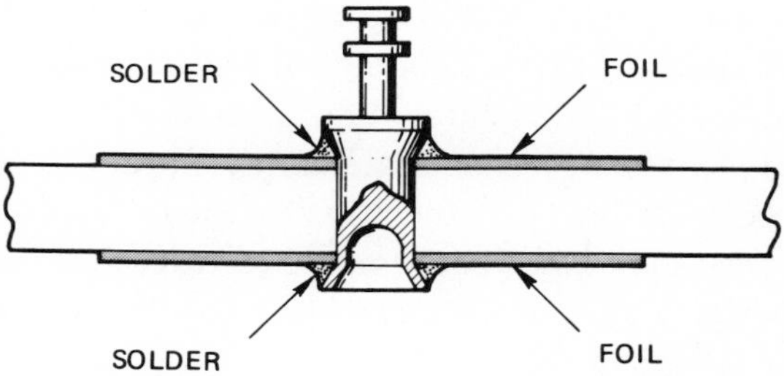

Figure 9-14 Stand-Off Terminal

9-8 TOLERANCES

Most dimensions on printed wiring boards are not critical. Since the edges are usually sheared, dimensions with respect to the edge are not too important and are not held in close tolerances. Hole centers are usually referenced to the reference holes. It is important to know what is easily achievable and not to design boards with tight tolerances on dimensions unless absolutely required. At microwave frequencies for examples, tolerances must be much tighter.

Easily achieved tolerances for most card dimensions are indicated in Table 9–5. With a little effort tighter tolerances can be held, but generally the values shown are good enough. The tolerance on punched holes is in addition to the springback mentioned in the preceding section. Since tolerance on front-to-back registration is ± 0.020 in., diameters of pads on opposite sides which are to be registered must be large enough so that a hole drilled penetrates both.

9-9 MOUNTING COMPONENTS

All components should be mounted on only one side of the board. On single-sided boards, parts are mounted on the side opposite to the conductors. Components should be arranged on the board so that any one of them may be installed or removed without disturbing any other part. (This may not be a requirement if the assembly is considered unrepairable.) Parts must not interfere with mounting holes nor be so close to the edge of the board as to interfere with slides, clasps, or other retaining devices.

The designer should consider thermal problems. Heat-producing elements such as tubes or power transistors must be insulated from the board to prevent damage to the laminate. A good rule to follow is the specification in MIL-STD-275 that any part dissipating one watt or more must be mounted so that the body of the part is not in direct contact with the board unless a thermal ground plane is used to dissipate sufficient heat to keep the temperature within acceptable limits. If there are thermally sensitive components on the board, they should be widely separated from those producing heat. On boards which are mounted vertically, heat producing components should be located near the top of the board.

Lightweight components may be supported by their leads, but heavier components must be mounted by clamps or other means to render them immobile. In general, parts weighing leass than $\frac{1}{4}$ oz per lead may be supported by their leads alone, but heavier components need auxiliary support. A component with two axial leads which is *self-supported*, that is, supported only by its own leads, should be in direct contact with the board, as shown in Fig. 9–15(a), unless it dissipates too much heat. However, on double-sided boards, the body of the component may bridge one or more conductors, in which case it may be enclosed in insulating sleeving. This shown in (b). Each component should be centered between mounting holes, so that the two lead lengths are approximately equal.

Parts should be located so that mounting holes fall on intersections of the modular grid. However, if a part has rigid mounting leads on the same side, spaced so they cannot all fall on grid intersections, then at least one such mounting hole should be on a grid intersection and the others dimensioned from the grid.

For horizontally mounted components with axial leads, the minimum horizontal length of each lead from the body of the component to the point at which it is bent toward its mounting hole should be 0.125 in. If a component has a bead or meniscus protruding from the body, as in some paper capacitors, the leads should extend at least 0.125 in. beyond the beads. Similarly,

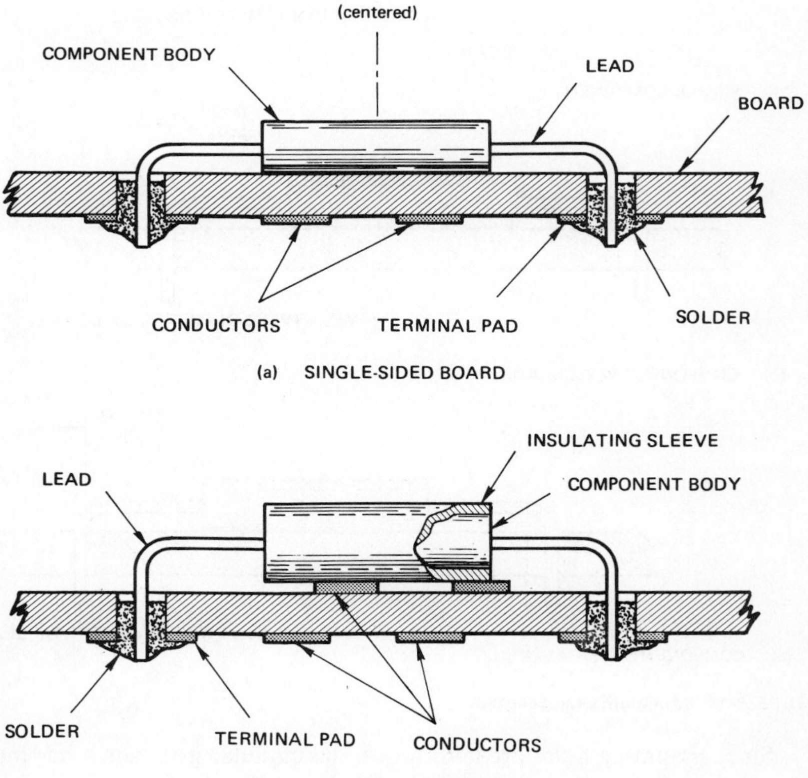

(a) SINGLE-SIDED BOARD

(b) INSULATED BRIDGE

FIGURE 9-15 Horizontal Mounting

in some tantalum capacitors, welds occur in the leads outside of the body. The leads then must extend at least 0.125 in. beyond these welds. These conditions are indicated in Fig. 9–16. The bend in the lead should have a minimum radius equal to twice the diameter of the lead. Although 0.125 in. is recommended as a minimum lead length, shorter lengths down to 0.060 in. are permissible.

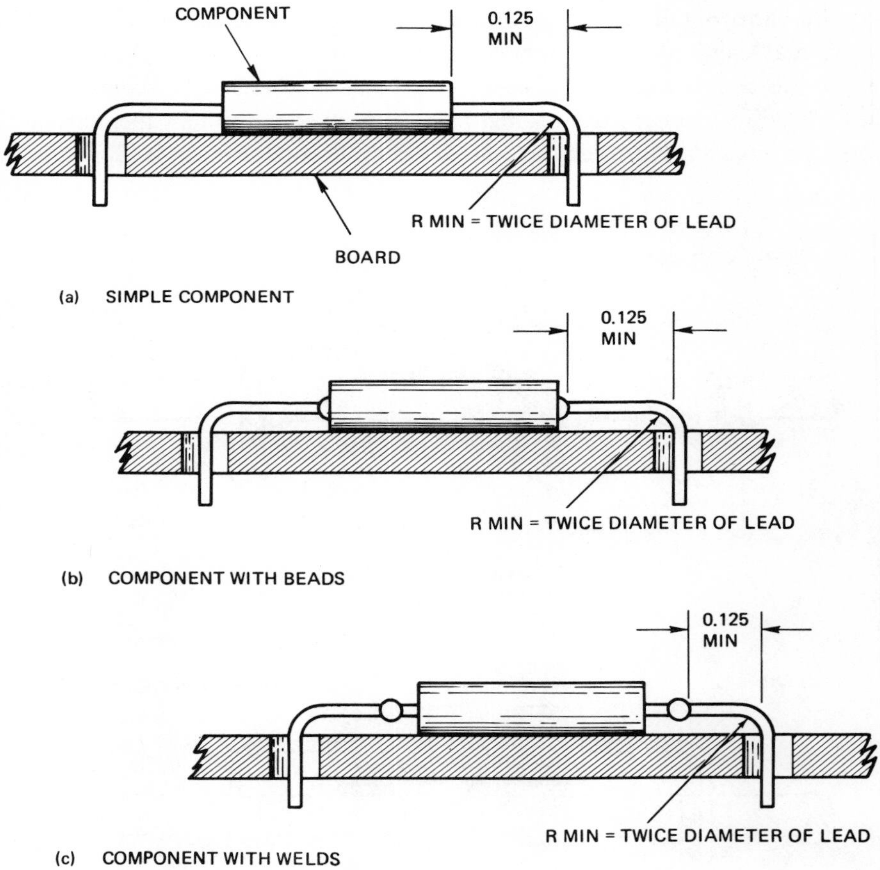

FIGURE 9-16 Minimum Lead Lengths

Since mounting holes are laid out on the modular grid which has intersections spaced 0.100 in. apart, the spacing of holes for a specific component will determine the lead length. For example, assume a resistor which is 0.562 in. is to be mounted. Allowing 0.125 in. for each lead makes a total length of

0.562 + 0.250 = 0.812 in. A small additional amount must be allowed for the bends. This is then increased to the next multiple of 0.100 in. or to 0.900 in. This is the minimum recommended spacing. If space is at a premium, lead lengths of 0.060 in. may be used, then the total length is 0.562 + 0.120 = 0.682 plus a few thousandths of an inch for bends. This may be increased to the next multiple of 0.050 in. or 0.700 in. However, bending the leads closer to the component body requires more care in handling. A list of some common components with their body lengths and minimum recommended spacing for mounting holes is given in Table 9–6. The body length for tantalum capacitors includes the welds on the leads, and for paper capacitors the beads at the end of the body.

Small parts, weighing less than $\frac{1}{2}$ oz and having two axial leads, may also be mounted vertically, as shown in Fig. 9–17. The top lead may be bent back to a mounting hole on the board as shown in (a) or it may be soldered to a soldering post or terminal post as shown in (b). As indicated, the bottom of the component must be separated from the board from 0.030 to 0.060 in.

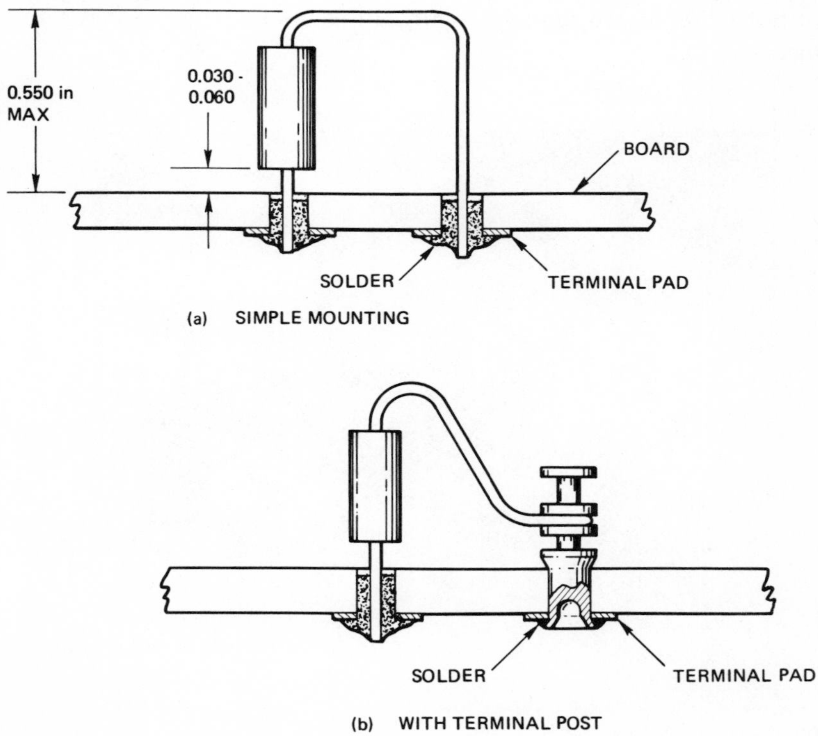

FIGURE 9-17 Vertical Mounting

The overall height of the mounting above the board should not exceed 0.550 in. In Fig. 9–17(a), the lead from the top of the component may be enclosed in a supporting sleeve. The minimum dimensions shown meet military specifications. Deviations may be permitted for non-military applications.

Semiconductor subassemblies, consisting of one or more transistors or diodes and additional components or conductors, may be packaged in separate cans or containers with all enclosed parts connected internally. Such a package may then be mounted on a printed wiring board along with other components.

Figure 9–18 shows the component side of a one-sided printed wiring board. Since the board material is translucent, the conductors and pads on the far side are visible in the photograph. Among the components are transistors, diodes, resistors, and capacitors. Figure 9–19 is a photograph of the conductor side of a PWB, showing more than a hundred solder connections, as well as "land" masses. Soldering is easily accomplished by placing the conductor side of the board on a hot solder bath. Solder adheres to the metal, but not to the board material. Thus, all connections are soldered simultaneously.

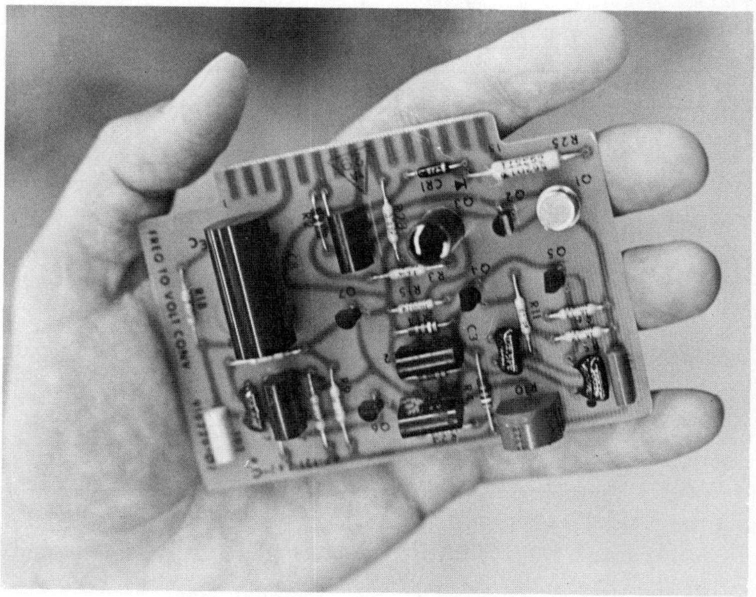

FIGURE 9-18 Component Side of PWB (Courtesy of Varian)

FIGURE 9-19 Conductor Side of PWB (Courtesy of Varian)

9-10 RF PROBLEMS

In circuits designed for dc applications, the positions of components and meandering of conductors are uncritical and need satisfy only mechanical or esthetic criteria. However, at alternating frequencies, even as low as the 60-hertz line frequency, there may be unwanted coupling between adjacent conductors, and stray reactances affect a circuit's performance. These effects increase as signal frequency is increased and must be considered in the design of printed wiring boards for radio frequencies.

In a PWB used only for dc circuits, the designer needs only a schematic diagram and can lay out parts and conductors as he wishes. For ac circuits, switching circuits, or logic circuits, the designer must work closely with the electrical engineer, since some experimentation is usually needed to determine the best design. It is rare that the first design works.

At radio frequencies, any length of conductor on the board is an inductance and two adjacent conductors, whether on the same side of the board or on opposite sides, present a capacitance to the current in the circuit. For conductors on opposite sides of the board, the capacitance between them is directly proportional to the dielectric constant of the material between them. Epoxy-glass laminates have a dielectric constant between 4 and 5, which means that the capacitance between conductors on opposite sides of the board is 4 to 5 times as much as it would be for the same spacing in air.

The bandwidth of amplifiers and oscillators is determined by the Q of the circuit and is limited by that part of the circuit having the lowest Q. A low Q results in widening the bandwidth. The epoxy-glass boards have a Q of 50, which is just barely satisfactory. Teflon or polytetrafluoroethylene has a Q greater than 100 and a dielectric constant of about 2 and is thus superior electrically to the epoxy-glass boards. However, they are more expensive, and, consequently, epoxy-glass boards are usually used if a suitable design can be determined.

Unwanted coupling is the biggest problem in laying out a PWB for radio frequencies. A common error is to ignore the ground return in a circuit. However, when two circuits are connected, some portion of the ground loop may be common to both, and coupling may occur in the shared loop, even though it is at ground potential. It is better to avoid common grounds.

Any conductor on the board must be thought of as a transmission line and is a potential source of radiation and coupling to another conductor or transmission line. Each line should be short and isolated from other lines. The board does *not* furnish isolation, but in fact it enhances coupling between conductors on opposite sides because of its dielectric constant. One method of discovering potential coupling points involves covering the rough layout with a piece of transparent material and marking on it every current loop in the circuit. Wherever these loops intersect, there is a possibility of coupling.

An additional problem is the stray capacitance which might exist between the PWB and the case when the electronics assembly is complete. Hence, even if a card works by itself on the bench, it should not be considered completed until it has been tested in its final assembly.

10

CHOICE OF COMPONENTS

The designer selects the components which are to be used in the final circuit. With the many hundreds of different components available, this is a difficult task. Not only must choices be made between, for example, a composition resistor and a wirewound resistor, but also between manufacturers of the same product, since there are differences here, too.

10-1 FACTORS AFFECTING CHOICE

Some characteristics considered in choosing a component are: size and weight, electrical limitations, environmental limitations, mechanical limitations, life, reliability, safety, and cost.

The relative importance of these factors is determined to a large extent by the marketplace. Cost is very important in an item designed for the competitive entertainment market, whereas reliability, safety, and life are more important in medical electronics.

Size and weight are important considerations for airborne equipment, space applications, and, to a lesser extent, for any portable equipment. The emphasis is on smaller, lighter equipment, which means smaller, lighter components. However, the whole assembly must be considered. If a lighter capacitor needs an additional clamp, for example, the added weight of the clamp may offset the savings in weight of the capacitor.

Electrical limitations include ratings for maximum voltage, maximum current, and maximum power. All components must be able to withstand expected currents and voltages without breakdown. This includes surge currents in starting and stopping equipment. Other electrical considerations include shielding for radio frequency interference and generation of noise.

Environmental limitations include effects of temperature, humidity, and pressure. Many components change drastically with changes of temperature and these components should not be used if the equipment is to be used in an environment that is subject to a large temperature change. Similarly, moisture causes changes in some components. Pressure does not usually affect a component, but it does affect breakdown voltages, so components must be selected to meet conditions of reduced pressure if required by the application.

Mechanical limitations include effects of shock and vibration. These are important considerations for most military or industrial equipment, although not too important for the entertainment market.

The useful *life* of the equipment depends upon the useful life of each of its components. Some components fail suddenly while others slowly change until they no longer meet specifications. This change is termed *aging*. *Stability* is the opposite of aging and is the ability to remain unchanged with time.

Reliability is related to life in that a part that fails is unreliable. The overall reliability of equipment depends on the failure rate of each component.

Safety of personnel during manufacture, maintenance, repair, and operation is always an important requirement. If a part fails, the equipment must be safe to handle. Since this is not always possible, critical components must be selected with an extra margin of safety in mind. For example, if a bleeder resistor fails, a high voltage could be present with the equipment turned off. This is a potential danger. Hence, bleeder resistors should be conservatively rated.

Cost is not always the most important consideration, but when everything else is equal, cost may be the deciding factor.

The task of selecting suitable components is simplified somewhat by past history. Problems which have been solved before by certain selection can be solved again in the same way. This does not mean that there is no room for innovation, but rather that the first step in the selection process should be consideration of what has gone before. Most companies usually standardize on a limited number of components, and a designer should try to select from these preferred parts, since they are usually readily available in company stock.

10-2 FIXED RESISTORS

Fixed resistors are made in three basic types: composition, film, and wirewound. They may be sealed for protection from moisture or unsealed.

Selection is dependent on resistance and tolerance, as well as the other factors mentioned in Sec. 10–1.

Wirewound and film resistors are available with a tolerance of 0.1%, whereas the best tolerance on the composition type is 5%. For stability, wirewound are best since they show no change with age. Film resistors may change ±10% over a long term, and composition resistors change ±20% if unsealed, ±15% if sealed. If size is important, composition resistors are smallest, and wirewound largest.

Frequency is a factor in selecting resistor types. Wirewound resistors are satisfactory at frequencies up to about 50 kHz. Above this frequency, the inductance of the wire and the capacitance between adjacent turns may affect operation of the circuit. Composition and film resistors may be used at higher frequencies, but composition resistors may be noisy and should not be used in sensitive circuits such as detectors or high-gain amplifiers. Film resistors have lowest noise.

Temperature and humidity problems are greatest with composition resistors, least with film types. Wirewound resistors are not bothered too much by humidity but do change with temperature.

Table 10–1, taken from MIL-STD-199A, lists many commercially available resistors of all three types. The first column contains the applicable military specification for each type, and the other columns have pertinent electrical and physical characteristics.

10-3 VARIABLE RESISTORS

Variable resistors are either wirewound or composition. As with fixed resistors, the wirewound have excellent long-term stability, whereas the composition type may change by ±20% with age. Wirewound resistors should not be used at frequencies above 50 kHz. Both types are affected adversely by moisture.

The sliding contact in composition resistors causes wear which changes the resistance after extended use. In wirewound resistors, the sliding contact causes noise as it moves from one wire to the next. Vibration may cause unwanted movement of the contact during operation, and some precautions are necessary to prevent this. If the resistor is to be varied only occasionally, a short, locked shaft with a slotted end can be used. Otherwise, a high-torque shaft can be used to limit motion. If a knob is used to turn the shaft, it should be small, since a large knob permits a high rotational torque which may damage the internal stop in the resistor.

Table 10–2, taken from MIL-STD-199A, lists commercially available variable resistors. The first column contains the applicable military specification for each type, and the other columns have pertinent electrical and physical characteristics.

10-4 CAPACITORS

A capacitor consists of two metallic plates separated by a nonconducting material. The nonconducting material is called a *dielectric* and may be air or other gas, solid, or liquid. The amount of capacitance is proportional to the area of the plates and inversely proportional to their separation. It is also dependent on the dielectric. The factor by which the material increases the capacitance over what it would be in air is called the *dielectric constant* of the material. The plates may be rigid and self-supporting or they may be made of metal foil or a layer of metal electroplated on opposite sides of a piece of solid dielectric material.

There are six basic types of capacitors in general use. They are distinguished by their dielectrics, which are:

1. Air
2. Vacuum
3. Glass and mica
4. Paper and plastic
5. Ceramic
6. Electrolyte

Each type has its own advantages and disadvantages.

Air and vacuum capacitors are affected by vibration, which may cause the spacing between the plates to vary or change permanently. Paper, plastic, or electrolytic capacitors may age due to chemical changes. Moisture accelerates these changes. However, glass, mica and ceramic capacitors have excellent long-term stability.

The electrical characteristics of some of the more common capacitors and their physical dimensions are given in MIL-STD-198B, which is available from the Government Printing Office in Washington D. C.

11

MICROWAVES

At microwave frequencies, wavelengths of the order of an inch or less present special problems in manufacture and design of equipment, since a dimension in the circuit may be an appreciable fraction of a wavelength. A slight change in spacing, which is permissible under normal tolerances, may have a deleterious effect on the performance of the circuit.

Tolerances must be tighter and in some cases are too tight for normal fabrication techniques. Special techniques have been developed to solve some of these problems, and these techniques should be incorporated into the design and called out on the drawings to eliminate the necessity of the shop having to reinvent solutions to problems which have already been solved. The designer should maintain or have available a catalog of *standard techniques* for problems already solved.

11-1 WAVEGUIDE ASSEMBLIES

As with most assemblies to be used at microwave frequencies, the major problem is holding tolerances. Brazing or dip soldering causes distortion which sometimes exceeds specified tolerance. If parts are machined to correct tolerances, which may be ± 0.001 in., it is almost a certainty the assembly will be out of tolerance after the parts are brazed together. A solution to the problem is to braze first and machine to tolerances later. For example, a piece of waveguide is needed with flanges spaced 2.000 ± 0.001 in., as shown in Fig. 11–1. The waveguide should be cut roughly to some longer dimension, perhaps $2\frac{1}{16}$ in. The flanges are brazed in place so that their external spacing exceeds the required 2.000 in., as shown in Fig. 11–2. The exact spacing is not at all critical. After brazing, the excess waveguide and the face of each flange are milled off so that the overall length meets the required specification of

2.000 $\pm$ 0.001 in. Since flanges are used for mechanical support only, their thicknesses are not critical.

If two waveguide assemblies must be the *same* length within very close tolerances, while the length itself is not critical, each piece is made separately up to final machining. Then the two are machined together to have the same overall length. The assembly of each, with loose tolerances, may be as indicated in Fig. 11–2, and then *matched machining* to length is specified. This final length may or may not have tight tolerances, depending on the requirements of the specific assembly.

Any deformation in a waveguide wall may affect electrical performance of a circuit. The closer the deformation is to the center-line of the broad wall of the waveguide, the greater is its effect. A dent in the broad wall or a burr inside the guide acts as an unwanted capacitor across the waveguide. Any distortion changes the electrical length.

If a hole is drilled in a waveguide wall, it can give rise to two problems. First, the pressure of the drill may cause a dimple in the wall, and secondly, there will be a burr inside the waveguide after drilling. If a metal block is placed inside the guide to fill the space before drilling, as shown in Fig. 11–3, both problems will be solved. This technique should be specified or referenced on the drawing in order to avoid scrapping or rework later.

Since dimensions are frequently in thousandths of an inch, it is usually impossible to show all detail on a full-size drawing. Thus, it is desirable to draw all small dimensions and spacings in enlarged detail drawings, so that the machinist can get a clearer picture of what he is making. Despite finer details and closer tolerances required in fabricating microwave assemblies, there is rarely a need for special tools. As with other conventional designs, inside square corners should be avoided.

Dimensions and tolerances for standard waveguides are shown in Table 11–1.

If a piece of standard extruded waveguide is to be used in the assembly, tolerances should not be called out which are closer than the tolerances on the *inside dimensions* of the waveguide.

Not all dimensions need tight tolerances, and loose tolerances should be specified whenever a design permits it. For example, the location of an iris may be critical in a *machined* waveguide and should then be specified with tight tolerances, but in a standard extruded product a variation in inside width would make the location of the iris uncritical. Thus in WR 90, the wide inside dimension of the waveguide can vary from 0.897 to 0.903 in., with a corresponding change in guide wavelength at 9 GHz from 1.922 to 1.907 in. This is a variation of 0.015 in. in guide wavelength, and iris locations need

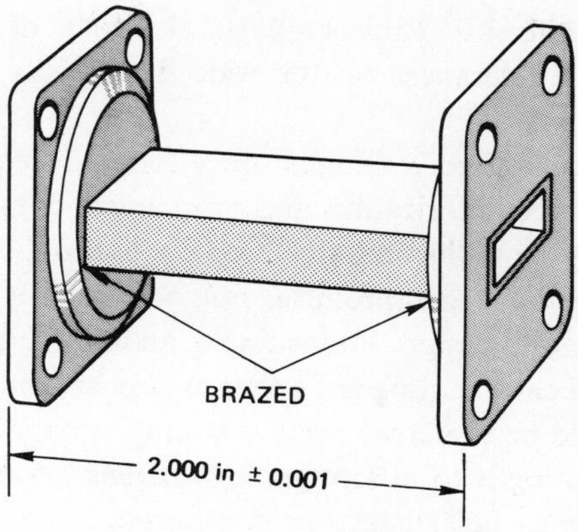

FIGURE 11-1 Flanged Waveguide

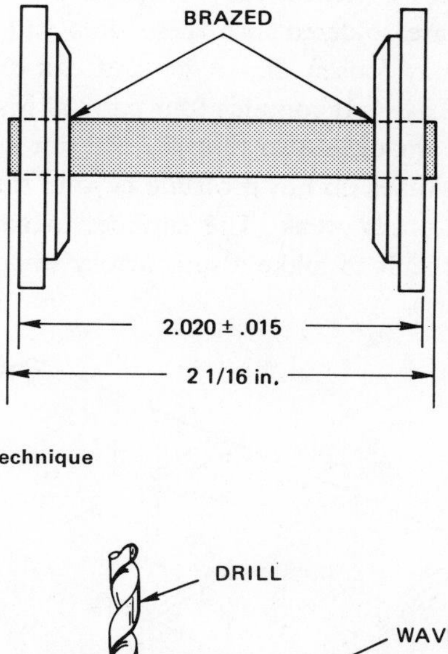

FIGURE 11-2 Assembly Technique

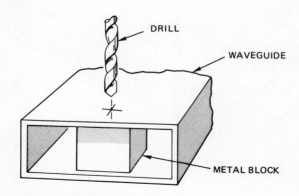

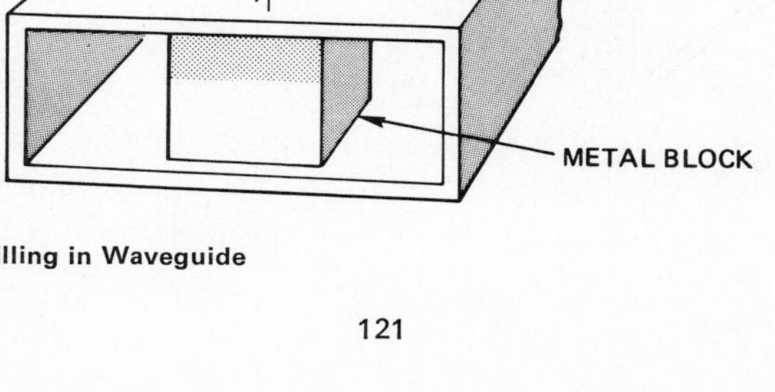

FIGURE 11-3 Drilling in Waveguide

not be held closer than this. Table 11–2 lists the range of guide wavelengths (λ_g) for extremes of tolerances on the wide dimensions of some common waveguides.

If two or more different designs are electrically equivalent, deciding factors should be ease of fabrication and mechanical considerations such as strength and ability to withstand shock and vibration. In some cases, the design will have to be a compromise, but occasionally all considerations point to one preferential design. For example, an inductive discontinuity may be needed, and this can be designed either as a post or a vane in the waveguide. A post should be preferred since it is simpler to fabricate than a slot, and a post adds strength to a waveguide, whereas a vane may weaken it. Electrically, either may be satisfactory in a particular job, although the two are not exactly equivalent.

A waveguide filter containing many vanes may not be able to withstand shock and vibration tests, because the waveguide is weakened by many slots. Even though vanes are soldered into these slots and can act as stiffeners, constant vibration may loosen the solder and cause a fracture. A typical filter is shown in Fig. 11–4. It contains four pairs of irises which are soldered in slots cut into the waveguide so that the exterior surface of the guide is smooth. That is, the vanes do not protrude beyond the guide. The design is esthetic but is structurally weak. The surfaces in contact are very small; consequently it is difficult to make a satisfactory solder joint.

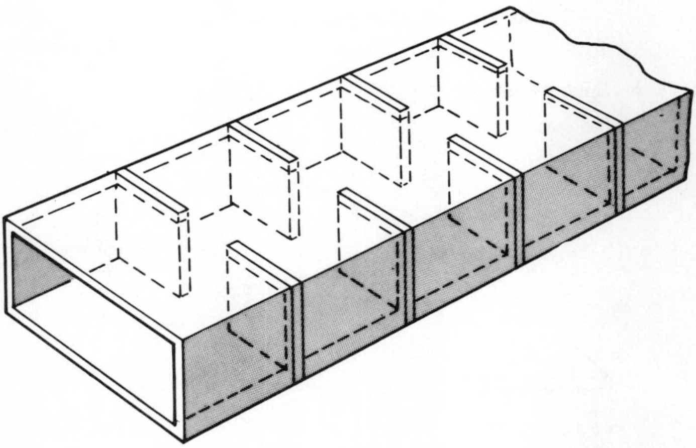

FIGURE 11-4 Waveguide Filter

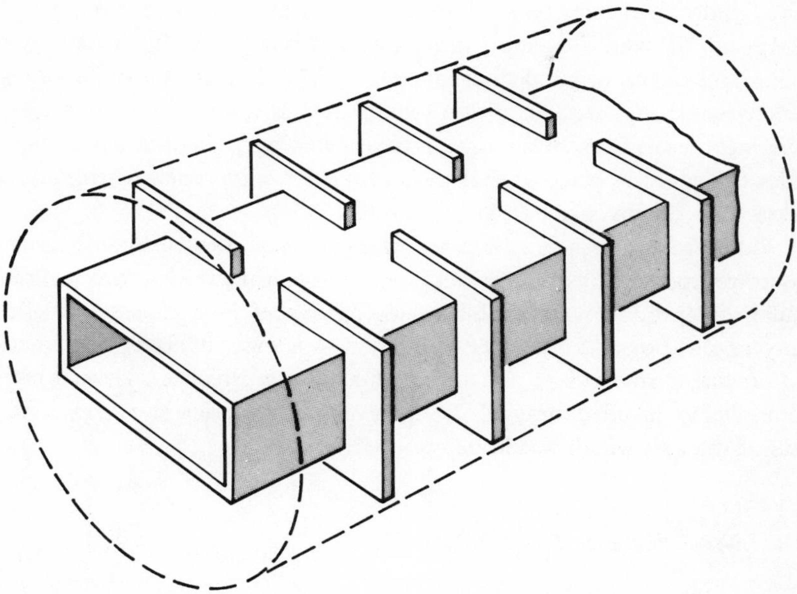

FIGURE 11-5 Extended Vanes for Strengthening

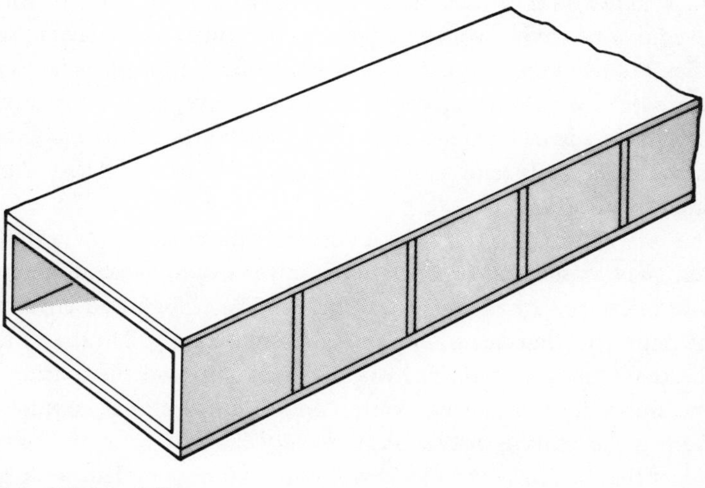

FIGURE 11-6 Reinforced Filter

A better design, if space permits, would have the vanes protrude beyond the waveguide as shown in Fig. 11–5. Since the vanes extend beyond the guide, solder fillets can be used at every seam to strengthen the joints. For eye-appeal, the whole assembly may be potted as indicated by the dotted lines in the figure. The original filter, shown in Fig. 11–4, may also be strengthened by soldering metallic plates to top and bottom. The solder used in fastening these plates must have a lower melting point than that used to solder the vanes in place so that the vanes are not loosened when plates are added. This design is shown in Fig. 11–6.

Once the design is fixed and a prototype tested, the waveguide assembly may be reproduced by casting. Precision casting, using the lost wax technique, results in reproducible assemblies with tolerances held closely enough for many applications. If the filter of Fig. 11–4 is cast, it needs no structural reinforcement since there are no soldered joints. Precision casting can be economically justified only if the quantities are sufficient to warrant the costs of the dies which would be required.

11-2 MACHINED WAVEGUIDES

In some applications, a standard extruded waveguide is not accurate enough. Typically the system may require a rectangular resonant cavity which must be reproducible with closer tolerances than the ± 0.003 in. allowed for WR 90, which would be used at the frequency of the system. In such cases, the waveguide or cavity would be machined from solid material. Since it is too difficult to machine the guide as a hollow tube, it is more usual to machine it in two blocks which fit together to form the waveguide. Alternatively, the guide may be made of a trough and cover, which are usually made of aluminum, since thick walls tend to make the assembly heavy. These designs are illustrated in Fig. 11–7.

The two halves of a split block construction must be accurately pinned together, since misalignment of the two halves would cause more error than the loose tolerances of extruded waveguide. The trough and cover arrangement is better, in that accurate pinning is unnecessary. However, the cover must be thick enough to lie flat without buckling, and the mating surfaces must be smoothly finished to assure continuous electrical contact.

During the milling operation, it is very difficult to avoid burrs at the sharp corners. In removing the burrs, corners are rounded, with the result that the waveguide does not have a continuous sidewall. This is shown in exaggerated form in Fig. 11–8. The rounded corners result in a lowering of

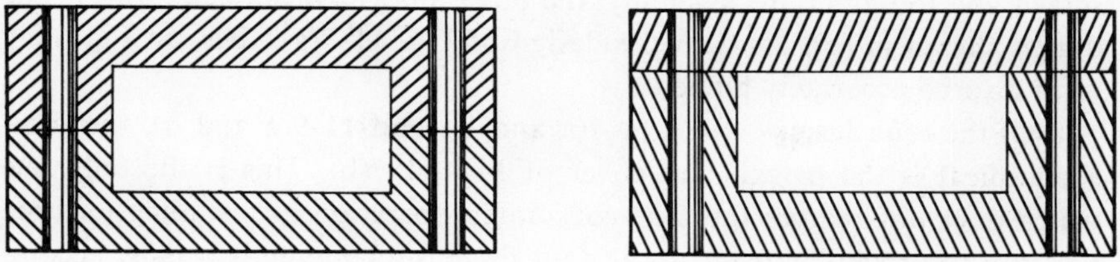

(a) SPLIT BLOCK (b) TROUGH AND COVER

FIGURE 11-7 Machined Waveguides

FIGURE 11-8 Rounded Corners

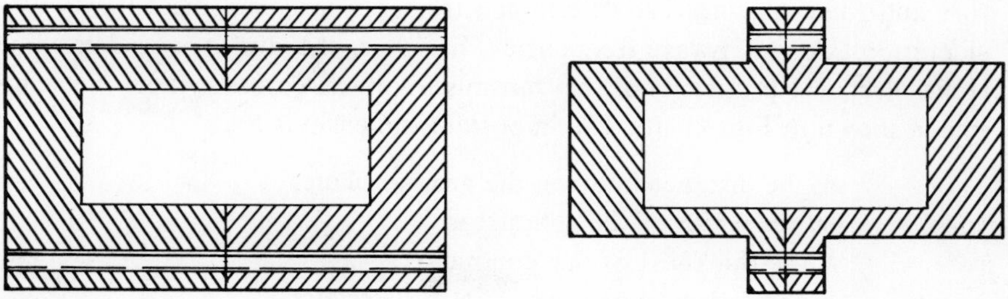

(a) SPLIT BLOCK (b) LIGHTENED

FIGURE 11-9 Split Parallel to Current Lines

Q if the waveguide is used for a cavity, but are not otherwise harmful unless the contact is disturbed and there is electrical leakage through the gap. This is possible since the break cuts lines of maximum current in the walls of the waveguide.

A method of splitting the guide without danger of leakage in case of poor contact involves making the split parallel to the lines of current flow so that the split does not cut any current lines. Possible arrangements are shown in Fig. 11–9. The design in Fig. 11–9(a) is a simple split block with the split occurring in the center of the broad walls. To avoid using long fastening screws and to lighten the assembly, the block may be machined down to the design shown in (b). As with the design of Fig. 11–7(a), these in Fig. 11–9 must also be accurately pinned.

Of the four designs, Figs. 11–7(a and b), and 11–9(a and b), the most economical is the trough and cover of Fig. 11–7(b). This is the only one which does not require the time-consuming operation of installing locating pins for accurate alignment. In all four designs, clamping is usually accomplished by using tapped holes in one half of the assembly and clearance holes in the other.

11-3 STRIPLINE

Stripline is a form of microwave transmission line, consisting of a conducting strip located midway between two ground planes. The conducting strip, or *conductor*, may have any cross section, such as round, square, or rectangular, although rectangular is most common. The center conductor may be supported by low-loss dielectric material, or it may be etched on copper-clad material similar to a printed wiring board.

At lower frequencies, when PWBs are used, the dimensions of conductors and their spacings are determined from physical and mechanical considerations. At microwave frequencies, however, the dimensions determine some electrical properties of the transmission line. A cross section of a stripline is shown in Fig. 11–10. The important dimensions are:

- b, the distance between the ground planes
- w, the width of the conductor
- t, the thickness of the conductor
- ϵ, the dielectric constant of the material

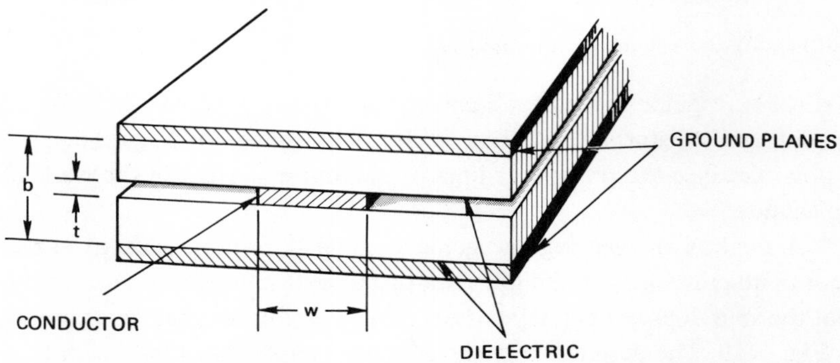

FIGURE 11-10 Stripline

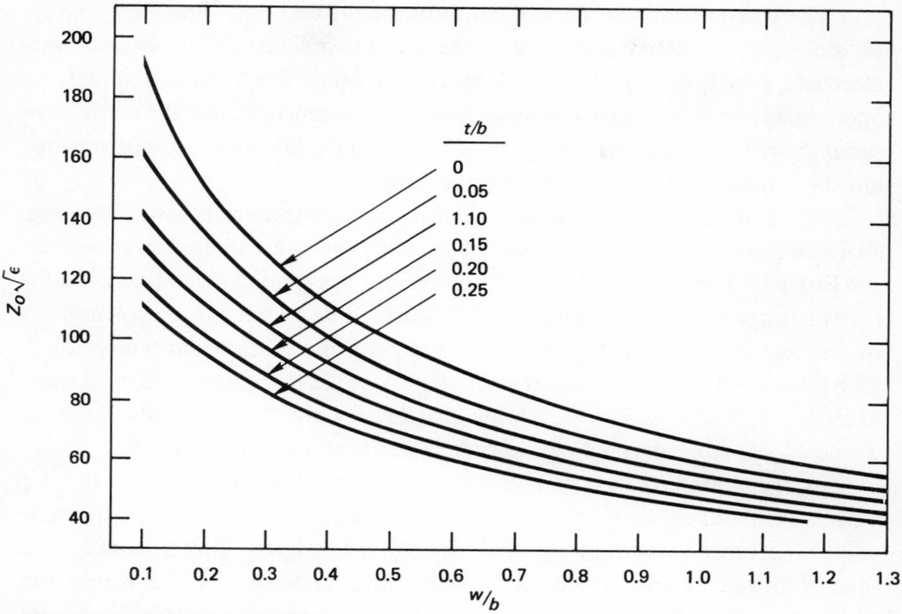

FIGURE 11-11 Impedance Chart

The characteristic impedance, Z_o, of the strip transmission line is determined from these four quantities. A graph of $Z_o(\epsilon)^{1/2}$ as a function of w/b for different values of t/b is shown in Fig. 11–11.

Example: For mechanical reasons, $b = 0.125$ in. and $t = 0.003$ in. The dielectric material has a dielectric constant of 2.56. Determine w for the line to have $Z_o = 50$.

Solution: $\epsilon = 2.56$, and therefore, $(\epsilon)^{1/2} = 1.6$.
$Z_o(\epsilon)^{1/2} = 80$.
$t/b = 0.024$, about halfway between 0 and 0.05.
From the chart, $w/b \approx 0.68$.
Therefore, $w \approx 0.085$ in.

If copper-clad dielectric material is used to make the stripline, the assembly can be very compact and lightweight. In Fig. 11–10, the assembly can be made from two boards, one of which has copper on both sides, and the other on one side. The board with copper on both sides has the conductor pattern etched on one side. Then the boards are placed together with the conductor in the middle, and the outer copper claddings act as the ground planes. The width of the ground planes should be at least four times the width of the center conductor.

The dielectric material must be uniform throughout, since any change of dielectric constant represents a change in characteristic impedance and electrical length of the line. This precludes using filled materials such as epoxy-glass or teflon-glass. Besides being homogeneous, the dielectric material should have low loss, good bonding strength, low moisture absorption, and be usable over a wide temperature range.

Manufacturers are continually developing new materials for microwave striplines. Some of the acceptable dielectrics and their dielectric constants are listed in Table 11–3. Textolite is a typical material designed specifically for this application. It is a General Electric trade name for polyphenylene oxide. Its dielectric constant is 2.56 at temperatures from room temperature to below −300°F, and it drops only slightly to 2.52 with increased temperature to well above 300° F. Manufacturers supply the materials listed in Table 11–3 with 1-oz or 2-oz copper cladding on one or both sides.

In the design shown in Fig. 11–10, it is evident that the conducting strip is not entirely surrounded by the dielectric, but instead has air on each edge. The error caused by these air spaces is negligible and is usually less than that caused by variations in width and thickness of the conductor within the tolerances of manufacture. However, to minimize this effect, the conductor should be as thin as practicable. The ground planes may be any convenient thickness.

Many microwave circuits involve coupling between two paths. Coupling between two conductors depends on the spacing between them, their orientation, and their cross sections. Two methods of coupling are shown in Fig. 11–12. In (a), both strips are located midway between the ground planes, and coupling is determined by the spacing between conductors. In (b), coupling is determined both by vertical spacing and by the dimensions of the strips. Since strip widths and spacings of 0.010 in. and less are sometimes required,

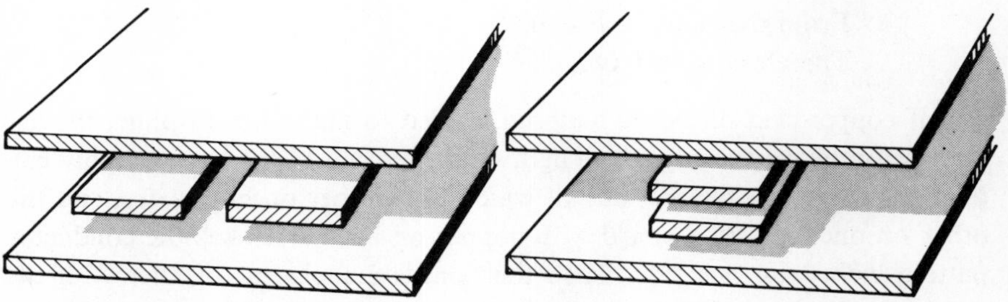

(a) SIDE BY SIDE (b) OVER - UNDER

FIGURE 11-12 Coupling

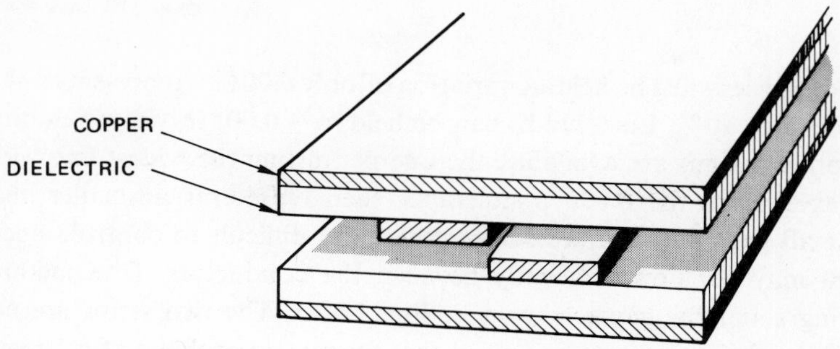

FIGURE 11-13 Narrow Gap Assembly

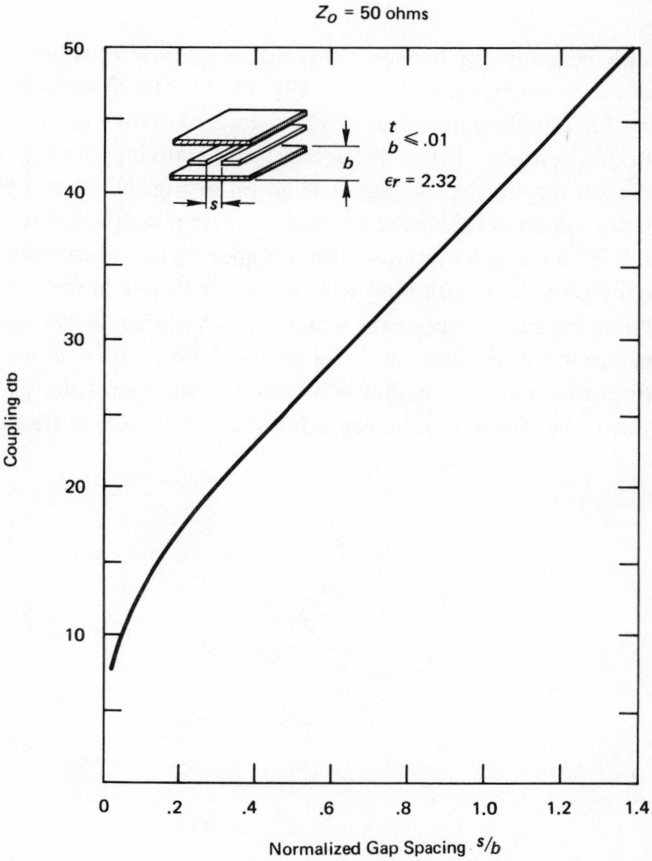

FIGURE 11-14 Coupling as a Function of Gap Spacing (Published in the 1970 *Microwave Engineers' Technical and Buyers' Guide* by Horizon House-Microwave, Inc., Courtesy of Harold Stinehelfer, Microwave Associates)

tight tolerances must be held. A variation of only 0.001 in. represents a change of more than 10%. Line widths can be held to ±0.001 in. during etching. If alternative designs are available, the one permitting the widest line width is preferable, since the 0.001 in. tolerance then represents a smaller change.

Small gaps between conductors are more difficult to control, since the etchant may not flow uniformly between the conductors. One method of attaining a narrow gap is shown in Fig. 11–13. The two strips are etched independently on separate boards so that there is no problem of getting etchant into a narrow gap. The two parts are brought together, and the gap tolerance is simply the tolerance of alignment. However, alignment may be more of a problem than controlling the etchant. The best solution is to avoid small gaps.

A graph of coupling between two adjacent quarter-wave strips as a function of their spacing, *s*, is shown in Fig. 11–14. The abscissa is s/b. For a fixed value of *b*, coupling increases as *s* is decreased. This means the value of coupling in db decreases. From the figure, loose couplings from about 50 db to 20 db are linear with the spacing. The graph of Fig. 11–14 was plotted for a line with polyolefin as the dielectric, and with strip widths cut for a 50-ohm line. If the line contains a material with a higher dielectric constant than the 2.32 value for polyolefin, coupling will be tighter (lower in db).

Most microwave components which are available in waveguide configurations can be duplicated in stripline to yield a much more compact, lightweight component. As with PWBs, once a successful design has been photographed, the circuit can be reproduced as often as required with great reliability.

12

MICROELECTRONICS

Reduction of physical size and weight of electronic equipment is a continuing design goal of the electronics industry. Specific requirements for miniaturization exist in airborne and space applications, but advantages are also evident in other areas, such as portable TV sets for the consumer, lightweight hearing aids and pacemakers, portable laboratory instruments, and minicomputers. *Microminiaturization* is the technique of attaining significant reductions in size and weight by using printed circuits, miniature components, and more efficient packaging. The relatively new technique of *microelectronics* is an advanced form of microminiaturization which is different in that *components* as well as their interconnections are "printed," thus making it possible to manufacture a whole circuit in space previously taken up by only a single component.

12-1 MICROELECTRONIC TECHNOLOGIES

In microelectronics, active and passive elements are combined inseparably on a single chip of dielectric material. Each chip is then a complete functional unit by itself and is called an *integrated circuit* or *IC*. The material on which the elements are printed or mounted is called a *substrate*. An integrated circuit may be as small as the head of a pin. The upper limit on size is determined by the number of elements in the circuit and economic considerations. If an error is made in fabrication, the whole circuit must be discarded. Also, it is uneconomical to have circuits which require too many different techniques in their manufacture. Sizes up to one-quarter of an inch square are common. ICs may be interconnected to form larger functional units or subassemblies.

There are two distinctly different technologies used to produce integrated circuits: *semiconductor* microelectronics and *thin-film* microelectronics. A thin film deposited on a substrate has electrical properties which depend on the thickness of the film. Passive components, such as resistors, capacitors and inductances, and their interconnections can be produced by thin films. Active components, such as transistors and diodes, are produced by semiconductor microelectronics. The semiconductor technology can also be used to produce some passive components.

A *thin-film integrated circuit* is a microelectronic circuit in which all elements are formed by depositing thin layers of material on a substrate or on the surface of a previously deposited layer. Interconnections are included in the films.

A *semiconductor integrated circuit* is produced by diffusing suitable materials in a semiconductor substrate to form active semiconductor junctions. Interconnections between active components are usually made with thin films.

A *hybrid integrated circuit* is a microelectronic circuit which was produced by more than one technology. Strictly speaking, a semiconductor integrated circuit which uses thin films for interconnections is a hybrid, although it is not usually so called since the interconnections are a very small part of the total circuit.

The term *thin-film hybrid circuit* is applied to an integrated circuit in which passive elements and interconnections are produced by thin-film deposition, but active components are formed separately and attached to the thin-film circuit mechanically.

The term *hybrid* may also apply to a combination of thin-film and lumped elements, semiconductor and lumped elements, or conventional assemblies in which some of the subassemblies are integrated circuits.

The physical dimensions of thin films and semiconductor elements are frequently less than a thousandth of an inch. It is customary to use the *mil*, equal to 0.001 in., as the unit of measurement. A *microinch*, equal to one-millionth of an inch, is also used. In the metric system, the unit used to measure small dimensions used in microelectronic circuits is the *micron*, which is one-millionth of a meter. Thus, 1 mil = 0.001 in. = 25.4 microns. The thickness of a very thin film which is much less than one mil thick is sometimes expressed in *angstrom units*, abbreviated Å. One micron equals 10,000 angstrom units.

12-2 PRACTICAL ASPECTS

The obvious advantage of microelectronic circuitry over conventional packaging is the tremendous reduction in size and weight even over printed wiring boards. This is demonstrated in Fig. 12–1, which illustrates progressive steps in the development of an electronic counter manufactured by Hewlett-Packard Corporation. At left in this photograph is the original vacuum-tube counter designed in 1952. In the center is the printed wiring board version built in 1960. At right is the integrated-circuit counter developed in 1967. This last consists of three integrated circuits which together occupy less room than the nixie tube used for display.

Although reduction of size and weight were primary goals in the development illustrated in Fig. 12–1, other benefits were obtained as well. With

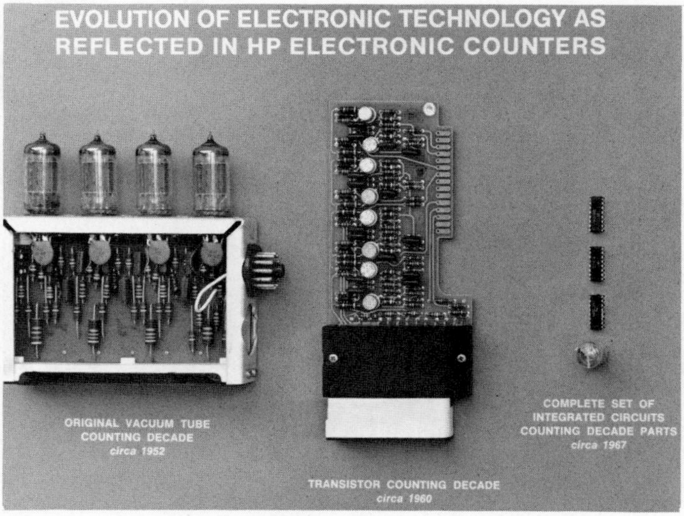

FIGURE 12-1 Evolution of Decade Counter (Courtesy of Hewlett-Packard)

each step in the development (from chassis to PWB, and from PWB to IC), speed of operation increased, reliability increased, and power consumption decreased.

Increase in reliability and decrease in power consumption are typical advantages of integrated circuits. There are two main reasons for increased reliability. First, the circuit (including components) is printed by a photographic process, so that once a satisfactory design is achieved, it may be reproduced over and over without possibility of human error. Secondly, interconnections among elements on an IC are also printed, so that failures due to poor connections are eliminated.

Despite its miniature size, an integrated circuit is frequently more complicated than many conventional circuits. In Fig. 12–2, a phase-locked loop

FIGURE 12-2 Integrated Circuit (Photo Courtesy of Signetics Corporation, a Subsidiary of Corning Glass Works)

linear IC is shown as the *small dot* in the center of the TO-100 can that is held in the fingers. The background is a greatly magnified photograph of the circuit as it would appear through a microscope, showing the many components and interconnections in this miniature chip.

Lack of flexibility of the IC is its major disadvantage. Each microelectronic chip is produced as a complete circuit, and design errors cannot be rectified except by discarding the IC and starting over. Retooling may be expensive.

12-3 MICROELECTRONIC TECHNIQUES

The design of a microelectronic circuit begins with the completion of a working breadboard model of the circuit. The designer must now translate the breadboard circuit into a set of layouts which will be used as patterns to make *masks*, which are used in fabricating the IC. Since the components in the circuit are not all the same, they are not all formed in a single step. Thus, the designer must prepare a separate layout for each group of components, and the sum of all of these layouts, when carefully aligned, represents the whole circuit. Typically, all thin-film resistors may be drawn on one layout, conductors on another, capacitors on a third, etc.

The designer draws the layout as much as 500 times full scale. This will be reduced photographically to make the needed masks. The artwork can be prepared on automatic machinery, and errors in dimensions will be reduced by the scale factor during photographic reduction. The artwork is photographed by a *step-and-repeat* process to produce many identical images on a glass photographic plate. The artwork may be reduced first, or it may be reduced during the step-and-repeat process.

The photographic plate is a *photographic mask* for preparing the IC. It may contain several hundred identical images which are subsequently transferred to a substrate. Thus the substrate, when completed, will contain several hundred identical integrated circuits. The substrate is then cut up into dice or chips, each of which is a separate IC.

The term, "mask," is also applied to two other steps in the semiconductor process—the *photoresist mask* and the *oxide mask*. These are illustrated in the simplified photoengraving process shown in Fig. 12–3. First the surface of a silicon wafer is oxidized as shown in (a). This oxide layer may be about 10,000 Å thick. Then a photoresist about 5000 Å thick is applied, as shown in (b). The photographic mask, with emulsion side next to the resist, is placed on the wafer, (c), and the assembly is exposed to ultraviolet light. In development unexposed parts of the photoresist are washed away, leaving a *photoresist mask*, shown in (d). This masks the oxide during etching. Figure 12–3(e) indicates the appearance of the wafer after etching has eaten away the unmasked portions of the oxide. Finally, the photoresist is removed. The *oxide mask* remaining, (f), permits diffusion of semiconductor impurities in selected areas.

Since a semiconductor junction exhibits resistance and capacitance, it is possible to fashion a resistor or capacitor in this way. Although a thin-film capacitor or resistor may be simpler than its semiconductor counterpart,

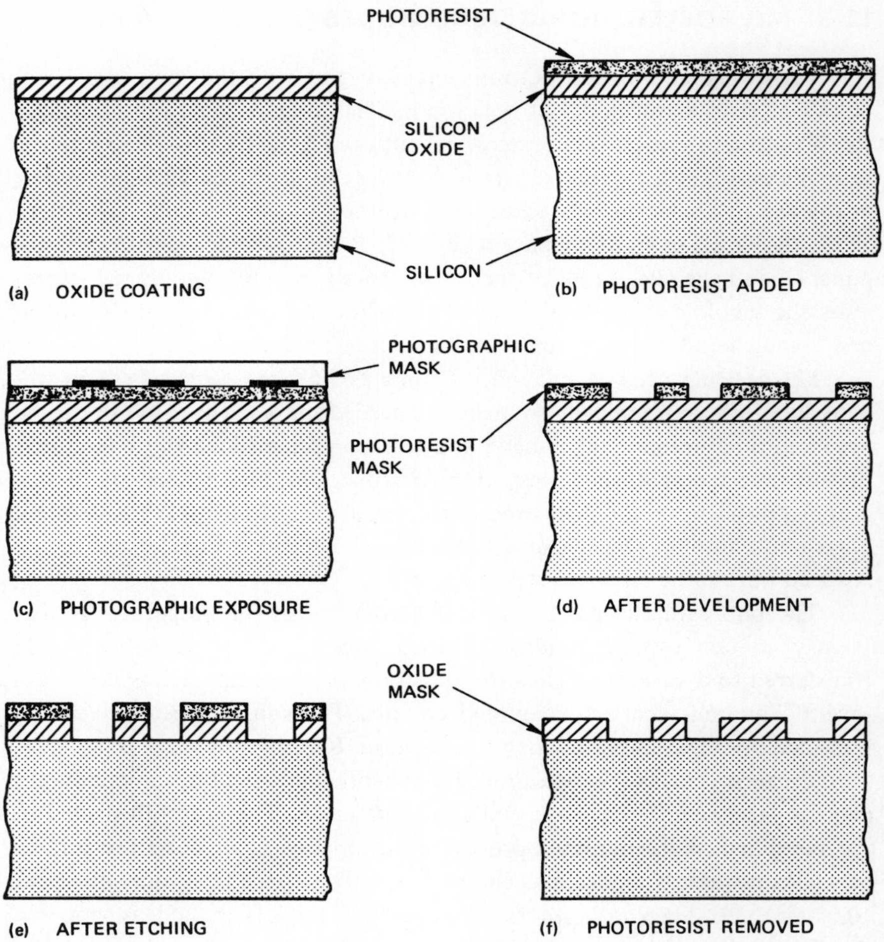

FIGURE 12-3 Simplified Photoengraving Process

it may be more economical to use the latter when other semiconductor components, such as transistors or diodes, are also being fabricated. With this technique all semiconductor elements can be processed together, whereas if thin-film components were also added, the two groups would require separate processing.

A thin-film resistor is simply a length of resistive metal deposited on the substrate. The resistance is linearly proportional to the resistivity of the metal and the length of the deposit. It is inversely proportional to the thickness and the width of the strip. This is shown in Fig. 12–4, where ρ is the resistivity of the material.

A thin-film capacitor, shown in Fig. 12–5, consists of two parallel plates of metal separated by a dielectric. A frequently used dielectric is silicon monoxide which is only 1000 Å thick. Capacitance is directly proportional to the area of overlap of the plates and is inversely proportional to the thickness of the dielectric. It is also proportional to the dielectric constant.

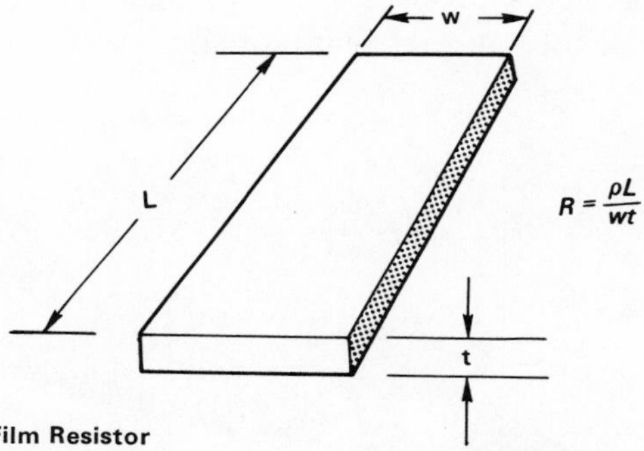

FIGURE 12-4 Thin-Film Resistor

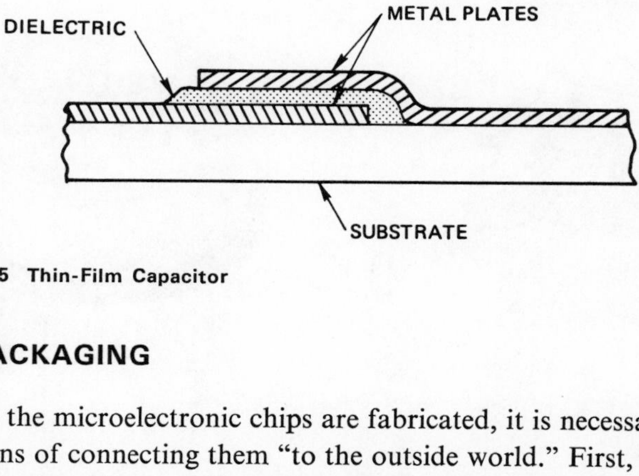

FIGURE 12-5 Thin-Film Capacitor

12-4 PACKAGING

After the microelectronic chips are fabricated, it is necessary to provide some means of connecting them "to the outside world." First, leads must be attached to the chip by soldering or welding. Leads may be in the form of wires less than half a mil thick, or in the form of metal ribbons as thin as 0.15 mil. After the leads are attached, the chip is packaged, and the leads are connected internally to external connections on the package.

One form of package is a rectangular *flat pack* enclosing the chip. Leads may be brought out from two opposite sides of the rectangle, or from all four sides if it is so desired. The package itself may be fabricated from metal, glass and metal, or it may be molded. The integrated circuits in Fig. 12–1 are packaged in flat packs.

A more economical package is a standard TO transistor can which comes in a variety of sizes. Although it is not as efficient a package as the flat pack and is restricted in that all the leads come off in one direction, its availability and low cost make it especially attractive. A microelectronic broadband amplifier, packaged in a TO-8 can, is shown in the photograph of Fig. 12–6. At left it is shown with cover removed to illustrate the method of mounting. The packaged circuit is shown at right. The pencil point in the photograph may be used as a reference for size.

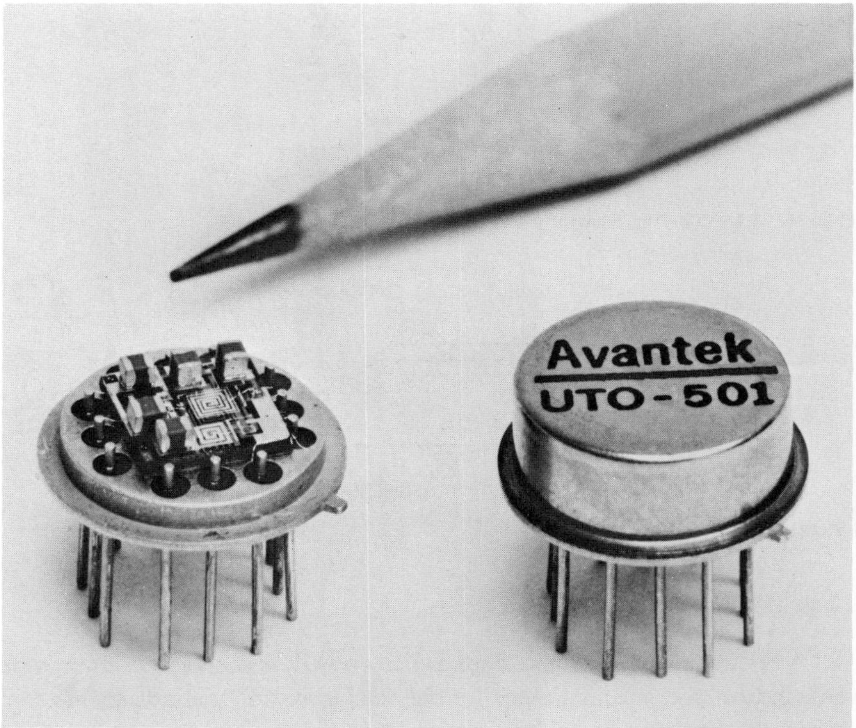

FIGURE 12-6 Broadband IC Amplifier (Courtesy of Avantek)

12-5 DESIGN CONSIDERATIONS

The designer of microelectronic circuits is concerned with reliability and cost; performance is only a secondary consideration. Thus, he must accept and, in fact, strive for the *lowest acceptable* performance which will be consistent with increased reliability and reduction of cost.

In preparing layouts, the designer must be concerned with the *compatibility* of different elements in the circuit. Wherever possible, elements requiring the same type of processing should be used, and the number of different processes should be minimized. As indicated in Sec. 12–3, this sometimes entails using semiconductor resistors with other semiconductor elements rather than the simpler thin-film resistors.

When several semiconductor components are fashioned on a single chip, *parasitic components* may be formed by action between adjacent elements. This is similar to stray capacitances in conventional circuits. For example, if two adjacent semiconductor components have *n*-layers in a *p*-substrate, the two *n*-layers and the substrate may form a parasitic transistor. Elements must be *isolated* to prevent parasitic elements from being formed.

In general, the thin-film approach permits faster fabrication time than the semiconductor approach. Also, there is no parasitic problem when using thin films. Semiconductor circuits may be smaller and lighter than thin films, but interconnections become more critical.

In designing a microelectronic *system*, the designer should use whatever standard components are available on the market, whenever possible. Beyond this, if additional circuits must be fabricated, they should not be so complicated that they have no practical use beyond the current system. It is better to use a few more chips and have each chip represent a circuit with many potential applications.

APPENDIX A

Reference Tables

TABLE 3-1 Main element in aluminum alloys

First Digit	Main Alloying Element
1	No added element, at least 99% aluminum
2	Copper
3	Manganese
4	Silicon
5	Magnesium
6	Magnesium and silicon
7	Zinc
8	All others

TABLE 3-2 Tempers of work-hardened alloys

Suffix	Meaning
F	As fabricated
O	Annealed (wrought alloys only)
H1	Strain-hardened
H2	Strain-hardened and annealed
H3	Strain-hardened and stabilized

TABLE 3-3 Tempers of thermally hardened alloys

Suffix	Meaning
T1	Cooled from an elevated temperature
T2	Annealed (cast alloys only)
T3	Heat-treated and cold-worked
T4	Cooled from elevated temperature, aged
T5	Heat-treated, aged
T6	Heat-treated, stabilized
T7	Heat-treated, worked, aged
T8	Heat-treated, aged, worked
T9	Cooled from elevated temperature, aged, worked

TABLE 3-4 Popular aluminum alloys

Alloy	Application
1100	Where conductivity and excellent corrosion resistance are required
2024	Structural; high-strength applications
3003	Moderate-strength plus workability
5052	Most popular for chassis, boxes, and brackets because of excellent weldability and good workability
6061	High-strength applications; good workability and corrosion resistance (stronger than 2024)
7075	Structural applications and for parts under high stress (strongest)

TABLE 3-5 Aluminum products

Alloy	Products Available
1100–0	All
2024–T3	Sheet, plate, tube
Alclad 2024–T3	Sheet
3003–0	All
3003–H14	Sheet, plate, tube
5052–H32	Sheet, plate, tube, wire
6061–T6	All
7075–T6	All
Alclad 7075–T6	Sheet, plate

TABLE 3-6 Available thickness of aluminum sheet and plate

Commonly Available Thickness	Weight	Commonly Available Thickness	Weight
(inches)	(lb./sq. ft.)	(inches)	(lb./sq. ft.)
0.006	0.086	0.160	2.30
0.008	0.115	0.190	2.74
0.010	0.144	0.250	3.60
0.012	0.173	0.313	4.51
0.020	0.288	0.375	5.40
0.025	0.360	0.500	7.20
0.032	0.461	0.625	9.00
0.040	0.576	0.750	10.80
0.050	0.720	1.000	14.40
0.063	0.907	2.000	28.80
0.071	1.02	3.000	43.20
0.080	1.15	4.000	57.60
0.090	1.30	5.000	72.00
0.100	1.44	6.000	86.40
0.125	1.80		

TABLE 3-7 Tolerances on thickness of aluminum sheet and plate

	1100, 3003			2024, 5052, 6061, 7075			
Thickness	Up to 18 in. Wide	18–36	36–54	Up to 18 in. Wide	18–36	36–48	48–54
0.006	.001	.001	.002	.001	.0015	.0025	.0025
0.008	.001	.0015	.002	.001	.0015	.0025	.0025
0.010	.001	.0015	.002	.001	.0015	.0025	.0025
0.012	.0015	.0015	.002	.0015	.0015	.0025	.0035
0.020	.0015	.002	.0025	.0015	.002	.0025	.0035
0.025	.0015	.002	.0025	.0015	.002	.0025	.0035
0.032	.002	.002	.0025	.002	.002	.0025	.004
0.040	.002	.0025	.003	.002	.0025	.003	.004
0.050	.0025	.003	.004	.0025	.003	.004	.005
0.063	.0025	.003	.004	.0025	.003	.004	.005
0.071	.0025	.003	.004	.003	.003	.004	.005
0.080	.003	.003	.004	.0035	.0035	.004	.005
0.090	.003	.003	.004	.0035	.0035	.004	.005
0.100	.0035	.004	.005	.004	.004	.005	.005
0.125	.0045	.0045	.005	.0045	.0045	.005	.005
0.160	.006	.006	.008	.006	.006	.008	.008
0.190	.007	.007	.009	.007	.007	.010	.010
0.250	.013	.013	.013	.013	.013	.013	.013
0.313	.013	.013	.013	.013	.013	.013	.013
0.375	.019	.019	.019	.019	.019	.019	.019
0.500	.025	.025	.025	.025	.025	.025	.025
0.625	.025	.025	.025	.025	.025	.025	.025
0.750	.030	.030	.030	.030	.030	.030	.030
1.000	.035	.035	.035	.035	.035	.035	.035
2.000	.060	.060	.060	.060	.060	.060	.060
3.000	.090	.090	.090	.090	.090	.090	.090
4.000	.110	.110	.110	.110	.110	.110	.110
5.000	.125	.125	.125	.125	.125	.125	.125
6.000	.135	.135	.135	.135	.135	.135	.135

TABLE 3-8 Tolerances on diameter of aluminum round stock

Diameter	Drawn Wire	Cold-Finished Rod
0–0.035 in.	0.0005 in.	
0.036–0.064	0.001	
0.065–0.374	0.0015	
0.375–0.500		0.0015 in.
0.501–1.000		0.002
1.001–1.500		0.0025
1.501–2.000		0.004
2.001–3.000		0.006
3.001–3.499		0.008
3.500–5.000		0.012

TABLE 3-9 Tolerances on distance across flats of aluminum (square, hexagonal, octagonal stock)

Distance Across Flats	Drawn Wire	Cold-Finished Bar
0–0.035 in.	0.001 in.	
0.036–0.064	0.0015	
0.065–0.374	0.002	
0.375–0.500		0.002 in.
0.501–1.000		0.0025
1.001–1.500		0.003
1.501–2.000		0.005
2.001–3.000		0.008

TABLE 3-10 Tolerance on thickness and width for aluminum rectangular stock

Thickness or Width	Tolerance on Thickness or Width
0–0.035 in.	0.001 in.
0.036–0.064	0.0015
0.065–0.500	0.002
0.501–1.000	0.0025
1.001–1.500	0.003
1.501–2.000	0.005
2.001–3.000	0.008

TABLE 3-11 Common thicknesses with tolerances of brass sheet

Sheet Thickness	Tolerance
0.013 in.	0.0013 in.
0.016	0.0015
0.20	0.0018
0.025	0.002
0.032	0.002
0.040	0.0025
0.050	0.003
0.063	0.003
0.080	0.0035
0.090	0.004
0.125	0.004
0.187	0.004
0.250	0.0045
0.312	0.005
0.375	0.005
0.500	0.005
0.625	0.007
0.750	0.007
1.000	0.009

TABLE 3-12 United States Standard Gauge for steel sheet and plate

Guage Number	Weight (lbs./sq. ft.)	Thickness (in.)
0000000	20.000	0.500
000000	18.750	0.469
00000	17.500	0.438
0000	16.250	0.406
000	15.000	0.375
00	13.750	0.344
0	12.500	0.312
1	11.250	0.281
2	10.620	0.266
3	10.000	0.250
4	9.375	0.234
5	8.750	0.219
6	8.125	0.203
7	7.500	0.188
8	6.875	0.172
9	6.250	0.156
10	5.625	0.141
11	5.000	0.125
12	4.375	0.109
13	3.750	0.0938
14	3.125	0.0781
15	2.812	0.0703
16	2.500	0.0625
17	2.250	0.0562
18	2.000	0.0500
19	1.750	0.0438
20	1.500	0.0375
21	1.375	0.0344
22	1.250	0.0312
23	1.125	0.0281
24	1.000	0.0250
25	0.875	0.0219
26	0.750	0.0188
27	0.688	0.0172
28	0.625	0.0156
29	0.562	0.0141
30	0.500	0.0125
31	0.438	0.0109
32	0.406	0.0106
33	0.375	0.0094
34	0.344	0.0086
35	0.312	0.0078
36	0.281	0.0070
37	0.266	0.0066
38	0.250	0.0062
39	0.234	0.0057
40	0.219	0.0054

TABLE 3-13 Physical and mechanical properties of aluminum and magnesium

	Metal	Ultimate Strength lb/in²	Yield Strength lb/in²	% Elongation in 2 in.	Shear lb/in²	Modulus of Elasticity Tension lb/in² × 10⁶	Density lb/in³	Minimum Melting Point °F
	Magnesium 10% Aluminum	20000	10000	1	10000	6.5	0.065	1100
Aluminium	1100–0	13000	5000	35	9000	10.0	0.098	1190
	2024–T3	70000	50000	18	41000	10.6	0.100	935
	Alclad 2024–T3	65000	45000	18	40000	10.6	0.100	935
	3003–0	16000	6000	30	11000	10.0	0.099	1190
	3003–H14	22000	21000	8	14000	10.0	0.099	1190
	5052–H32	33000	28000	12	20000	10.2	0.097	1100
	6061–T6	45000	40000	12	30000	10.0	0.098	1080
	7075–T6	83000	73000	11	48000	10.4	0.101	890
	Alclad 7075–T6	76000	67000	11	46000	10.4	0.101	890

TABLE 3-14 Physical and mechanical properties of steels

	Metal	Ultimate Strength lb/in²	Yield Strength lb/in²	% Elongation in 2 in.	Modules of Elasticity lb/in² × 10⁶	Density lb/cu. in.	Minimum Melting Point °F
	Cold-Rolled Low-Carbon	84000	76000	18	29.5	0.28	2800
Stainless	201	115000	55000	55	28.6	0.29	2550
	202	105000	55000	55	28.6	0.29	2550
	301	110000	40000	60	28.0	0.29	2550
	302	90000	40000	50	28.0	0.29	2550
	304	84000	42000	55	28.0	0.29	2550
	410	70000	45000	25	29.0	0.28	2700
	420	95000	50000	20	29.0	0.28	2650
	430	75000	50000	25	29.0	0.28	2600

TABLE 3-15 Characteristics of aluminum alloys

Alloy	Corrosion Resistance	Bending	Spinning	Machining	Resistance Welding	Arc Welding	Gas Welding	Soldering (after Plating)	Brazing
1100–0	A	A	A	B	B	A	A	C	A
2024–T3	D	D	D	B	B	B	C	C	D
Alclad 2024–T3	B	D	D	B	B	B	B	C	B
3003–0	A	B	B	B	A	A	A	C	A
3003–H14	A	C	D	B	A	A	A	C	A
5052–H32	A	B	C	B	A	A	A	C	D
6061–T6	B	B	D	C	A	A	A	C	A
7075–T6	C	D	D	D	B	C	D	C	D
Alclad 7075–T6	B	D	D	D	B	C	D	C	D

TABLE 3-16 Characteristics of copper alloys

Alloy	Corrosion Resistance	Cold Working	Hot Working	Brazing	Soldering	Resistance Welding	Arc Welding	Gas Welding
99.9% Copper	A	A	A	A	A	D	C	B
95% Brass	A	A	B	A	A	D	C	B
65% Brass (Yellow-Brass)	C	A	D	A	A	B	C	C
Phosphor Bronze 5%	A	A	D	A	A	B	B	B
Phosphor Bronze 8%	B	B	D	A	A	B	B	B
Nickel Silver 65–15	C	A	D	A	A	B	D	C

TABLE 3-17 Characteristics of steels

	Alloy	Corrosion Resistance	Hot Working	Cold Working	Spinning	Soldering	Brazing	Welding
Low-Carbon Steel		D	B	C	C	B	B	B
Stainless	201	B	B	B	B	B	B	A
	202	B	B	B	B	B	B	A
	301	B	B	B	B	B	B	A
	302	B	B	B	B	B	B	A
	304	B	B	B	B	B	B	A
	410	C	B	B	C	B	B	C
	420	C	B	C	D	B	B	C
	430	C	B	B	B	B	B	C

TABLE 4-1 Minimum bend radii

	Metal	Thickness (T) in Inches						
		$\frac{1}{32}$	$\frac{1}{16}$	$\frac{1}{8}$	$\frac{3}{16}$	$\frac{1}{4}$	$\frac{3}{8}$	$\frac{1}{2}$
Aluminum Alloys	1100–0	1T	1T	1T	1T	1T	1T	2T
	2024–T3	4T	5T	6T	6T	7T	8T	9T
	Alclad 2024–T3	3T	4T	5T	5T	6T	7T	8T
	3003–0	1T	1T	1T	1T	1T	1T	2T
	3003–H14	1T	1T	1T	1T	$1\frac{1}{2}$T	$2\frac{1}{2}$T	3T
	5052–H32	1T	1T	1T	1T	$1\frac{1}{2}$T	2T	$2\frac{1}{2}$T
	6061–T6	$1\frac{1}{2}$T	2T	3T	4T	4T	$5\frac{1}{2}$T	6T
	7075–T6	5T	6T	7T	7T	10T	11T	12T
	Alclad 7075–T6	4T	5T	6T	6T	9T	10T	11T
Copper Alloys	Copper	1T	1T	1T	1T	$1\frac{1}{2}$T	$1\frac{1}{2}$T	2T
	95%	1T	1T	1T	1T	$1\frac{1}{2}$T	$1\frac{1}{2}$T	2T
	65% Brass	1T	1T	1T	1T	1T	1T	1T
	Phosphor Bronze 5%	1T	$1\frac{1}{2}$T	2T	2T	$2\frac{1}{2}$T	3T	3T
	Phosphor Bronze 8%	1T	1T	$1\frac{1}{2}$T	2T	$2\frac{1}{2}$T	3T	3T
	Nickel-Silver 65–15	$1\frac{1}{2}$T	2T	$2\frac{1}{2}$T	3T	4T	5T	6T
Steels	Low-Carbon Steel	1T	1T	1T	1T	1T	1T	1T
	Stainless 200–300 Ser.	1T	1T	1T	1T	1T	1T	1T
	Stainless 400 Series	1T	1T	2T	2T	2T	2T	2T

TABLE 5-1 Strengths of solders

Solder	% Lead	% Tin	Tensile Strength PSI	Shear Strength PSI
80–20	80	20	4940	3300
70–30	70	30	5390	4000
60–40	60	40	5660	4500
50–50	50	50	5800	5120
40–60	40	60	6400	5400
Eutectic	38	62	6500	5500
30–70	30	70	5400	5000
20–80	20	80	2900	4250

TABLE 5-2 Weldability of metals by spot welding

	Aluminum	Carbon Steel	Stainless Steel	Brass	Copper	Nickel	Phosphor Bronze	Tin Plate	Zinc
Aluminum	B	D	X	D	X	D	C	C	C
Carbon Steel	D	A	A	D	X	C	C	B	X
Stainless Steel	X	A	A	X	X	C	D	B	X
Brass	D	D	X	C	D	C	C	D	X
Copper	X	X	X	D	X	D	C	X	X
Nickel	D	C	C	C	D	A	C	C	X
Phosphor Bronze	C	C	D	C	C	C	B	D	D
Tin Plate	C	B	B	D	X	C	D	C	C
Zinc	C	X	X	X	X	X	D	C	C

RATING

A Excellent
B Good
C Fair
D Poor
X Not Weldable

TABLE 5-3 Spot weld shear strength (pounds per spot)

Thickness of Thinner Sheet	Aluminum Alloys			
	1100–0 3003–0	3003–H14	5052–H32 6061–T6	2024–T3 7075–T6
0.016 in.	40	55	80	85
0.020	60	80	105	110
0.025	90	115	140	150
0.032	130	170	190	210
0.040	180	240	250	275
0.050	240	330	350	385
0.063	320	450	500	550
0.080	425	620	700	840
0.100	500	730	850	1050
0.125	630	840	1300	1700

TABLE 5-4 Spot weld dimensions for aluminum

Thickness of Thinner Piece	Minimum Overlap	Weld Spacing	Minimum Edge Distance
0.016 in.	3/8 in.	5/16 in.	3/16 in.
0.020	3/8	3/8	3/16
0.025	7/16	3/8	7/32
0.032	1/2	3/8	1/4
0.040	1/2	7/16	1/4
0.050	5/8	1/2	5/16
0.063	3/4	1/2	3/8
0.080	3/4	5/8	3/8
0.100	7/8	3/4	7/16
0.125	1.0	1.0	1/2

TABLE 5-5 Spot weld data for low-carbon steel

Thickness of Thinner Piece	Minimum Overlap	Weld Spacing	Minimum Edge Distance	Minimum Shear Strength
0.016 in.	3/8 in.	1/4 in.	1/4 in.	250 (lbs/spot)
0.020	7/16	3/8	1/4	325
0.032	7/16	1/2	5/16	650
0.040	1/2	3/4	3/8	950
0.050	9/16	7/8	7/16	1400
0.063	5/8	1	1/2	2000
0.080	11/16	1 1/4	1/2	2750
0.100	3/4	1 1/2	5/8	4000
0.125	7/8	1 3/4	3/4	5000

TABLE 5-6 Spot weld dimensions for magnesium

Thickness of Thinner Piece	Weld Spacing	Edge Distance
0.020 in.	1/2 in.	1/4 in.
0.032	9/16	1/4
0.040	3/4	1/4
0.050	3/4	5/16
0.063	1	3/8
0.080	1 1/4	7/16
0.100	1 1/4	1/2
0.125	1 1/2	5/16

TABLE 5-7 Riveting head allowance

Rivet Diameter	Round Upset Head Fig. 5–5(a) or 5–5(b)	Countersunk Upset Head Fig. 5–5(c)
$\frac{2}{32}$ in.	0.094 in.	0.062 in.
$\frac{3}{32}$	0.141	0.094
$\frac{4}{32}$	0.188	0.109
$\frac{5}{32}$	0.234	0.125
$\frac{6}{32}$	0.281	0.141

TABLE 5-8 Upset head dimensions

Rivet Diameter	Maximum Head Diameter	Maximum Height
$\frac{2}{32}$ in.	0.10 in.	0.040 in.
$\frac{3}{32}$	0.15	0.056
$\frac{4}{32}$	0.20	0.075
$\frac{5}{32}$	0.25	0.094
$\frac{6}{32}$	0.30	0.113
$\frac{7}{32}$	0.35	0.130
$\frac{8}{32}$	0.40	0.150
$\frac{9}{32}$	0.45	0.168
$\frac{10}{32}$	0.50	0.187
$\frac{11}{32}$	0.55	0.205
$\frac{12}{32}$	0.60	0.225
$\frac{13}{32}$	0.65	0.243

TABLE 5-9 Rivet clearance holes

Rivet Diameter	Diameter of Clearance Hole	Drill Size
$\frac{2}{32}$ in.	0.067 in.	51
$\frac{3}{32}$	0.096	41
$\frac{4}{32}$	0.128	30
$\frac{5}{32}$	0.159	21
$\frac{6}{32}$	0.191	11
$\frac{7}{32}$	0.228	1
$\frac{8}{32}$	0.257	F
$\frac{10}{32}$	0.323	P
$\frac{12}{32}$	0.383	W

TABLE 6-1 Standard drill sizes

Drill Size	Nominal Diameter	Drill Size	Nominal Diameter	Drill Size	Nominal Diameter	Drill Size	Nominal Diameter	Drill Size	Nominal Diameter
80	.0135	55	.0520	31	.1200	8	.1990	$\frac{5}{16}$	.3125
79	.0145	54	.0550	$\frac{1}{8}$	.1250	7	.2010	O	.3160
$\frac{1}{64}$	.0156	53	.0595	30	.1285	$\frac{13}{64}$	.2031	P	.3230
78	.0160	$\frac{1}{16}$	.0625	29	.1360	6	.2040	$\frac{21}{64}$	.3281
77	.0180	52	.0635	28	.1405	5	.2055	Q	.3320
76	.0200	51	.0670	$\frac{9}{64}$	.1406	4	.2090	R	.3390
75	.0210	50	.0700	27	.1440	3	.2130	$\frac{11}{32}$	.3437
74	.0225	49	.0730	26	.1470	$\frac{7}{32}$	.2187	S	.3480
73	.0240	48	.0760	25	.1495	2	.2210	T	.3580
72	.0250	$\frac{5}{64}$	.0781	24	.1520	1	.2280	$\frac{23}{64}$	.3594
71	.0260	47	.0785	23	.1540	A	.2340	U	.3680
70	.0280	46	.0810	$\frac{5}{32}$	.1562	$\frac{15}{64}$	.2344	$\frac{3}{8}$	.3750
69	.0292	45	.0820	22	.1570	B	.2380	V	.3770
68	.0310	44	.0860	21	.1590	C	.2420	W	.3860
$\frac{1}{32}$	.0312	43	.0890	20	.1610	D	.2460	$\frac{25}{64}$	.3906
67	.0320	42	.0935	19	.1660	E $\frac{1}{4}$	.2500	X	.3970
66	.0330	$\frac{3}{32}$	.0937	18	.1695	F	.2570	Y	.4040
65	.0350	41	.0960	$\frac{11}{64}$	.1719	G	.2610	$\frac{13}{32}$	.4062
64	.0360	40	.0980	17	.1730	$\frac{17}{64}$	.2656	Z	.4130
63	.0370	39	.0995	16	.1770	H	.2660	$\frac{27}{64}$	.4219
62	.0380	38	.1015	15	.1800	I	.2720	$\frac{7}{16}$	.4375
61	.0390	37	.1040	14	.1820	J	.2770	$\frac{29}{64}$	.4531
60	.0400	36	.1065	13	.1850	K	.2810	$\frac{15}{32}$	.4687
59	.0410	$\frac{7}{64}$	.1094	$\frac{3}{16}$	.1875	$\frac{9}{32}$	.2812	$\frac{31}{64}$	.4844
58	.0420	35	.1100	12	.1890	L	.2990	$\frac{1}{2}$	.5000
57	.0430	34	.1110	11	.1910	M	.2950		
56	.0465	33	.1130	10	.1935	$\frac{19}{64}$	.2969		
$\frac{3}{64}$	.0469	32	.1160	9	.1960	N	.3020		

TABLE 6-2 Tolerances on hole sizes

Drill Sizes	Minus	Plus
#80 to #61	0.0005	0.004
#60 to $\frac{1}{8}$	0.001	0.005
$\frac{1}{8}$ to $\frac{1}{4}$	0.001	0.006
$\frac{1}{4}$ to $\frac{3}{8}$	0.002	0.007
$\frac{3}{8}$ to $\frac{1}{2}$	0.002	0.008

TABLE 6-3 Center drills

Drill Number	Small Diameter	Large Diameter
11	0.047 in.	$\frac{1}{8}$ in.
12	0.062	$\frac{3}{16}$
2	0.078	$\frac{3}{16}$
13	0.094	$\frac{1}{4}$
4	0.125	$\frac{5}{16}$
15	0.156	$\frac{7}{16}$
5	0.188	$\frac{7}{16}$
6	0.219	$\frac{1}{2}$
7	0.250	$\frac{5}{8}$
8	0.312	$\frac{3}{4}$

TABLE 6-4 Tap and clearance drills

Tap Size		Diameter	Tap Drill		Clearance Drill		Clearance
NC	NF		Size	Diameter	Size	Diameter	
	0–80	0.060	$\frac{3}{64}$	0.0469	51	0.067	0.007
1–64		0.073	53	0.0595	47	0.078	0.005
	1–72		53	0.0595			
2–56		0.086	50	0.0700	42	0.094	0.006
	2–64		50	0.0700			
3–48		0.099	47	0.0785	36	0.106	0.007
	3–56		45	0.0820			
4–40		0.112	43	0.0890	31	0.120	0.008
	4–48		42	0.0935			
5–40		0.125	38	0.1015	29	0.136	0.011
	5–44		37	0.1040			
6–32		0.138	36	0.1065	25	0.150	0.012
	6–40		33	0.1130			
8–32		0.164	29	0.1360	16	0.177	0.013
	8–36		29	0.1360			
10–24		0.190	25	0.1495	6	0.204	0.014
	10–32		21	0.1590			
12–24		0.216	16	0.1770	1	0.228	0.012
	12–28		14	0.1820			
$\frac{1}{4}$–20		0.250	7	0.2010	H	0.266	0.016
	$\frac{1}{4}$–28		3	0.2030			
$\frac{5}{16}$–18		0.312	F	0.257	$\frac{21}{64}$	0.328	0.016
	$\frac{5}{16}$–24		I	0.272			
$\frac{3}{8}$–16		0.375	$\frac{5}{16}$	0.312	$\frac{25}{64}$	0.391	0.016
	$\frac{3}{8}$–24		Q	0.332			
$\frac{7}{16}$–14		0.438	U	0.368	$\frac{29}{64}$	0.453	0.015
	$\frac{7}{16}$–20		$\frac{25}{64}$	0.391			
$\frac{1}{2}$–13		0.500	$\frac{27}{64}$	0.422	$\frac{33}{64}$	0.516	0.016
	$\frac{1}{2}$–20		$\frac{29}{64}$	0.453			

TABLE 7-1 Galvanic series

Metal	Anodic Index (0.01 volt)
Gold, Platinum	0
Rhodium	10
Silver	15
Nickel	30
Copper	35
Brass	40
Stainless Steel	50
Chromium	60
Tin-Plate, Tin-Lead Solder	65
Iron	85
Aluminum Alloys	90
Cadmium	95
Zinc	125
Magnesium	175

TABLE 7-2 Relative corrosion resistance of aluminum alloys

Alloy	Normal Atmosphere	Industrial Atmosphere	Marine Atmosphere
1100–0	A	B	B
2024–T3	B	C	D
Alclad 2024–T3	A	A	B
3003–0	A	B	B
3003–H14	A	B	B
5052–H32	A	A	A
6061–T6	A	B	B
7075–T6	B	C	D
Alclad 7075–T6	A	B	C

RATING

A Excellent
B Good
C Fair
D Poor

TABLE 8-1 Screw dimensions (maximum)

Screw Number	Diameter	Round Head		Filister Head		Pan Head		Flat Head 100°	
		D	H	D	H	D	H	D	H
0	0.060	0.113	0.053	0.096	0.059	0.116	0.044	0.119	0.025
1	0.073	0.138	0.061	0.118	0.071	0.142	0.053	0.149	0.030
2	0.086	0.162	0.069	0.140	0.083	0.167	0.062	0.172	0.036
4	0.112	0.211	0.086	0.183	0.107	0.219	0.080	0.225	0.049
6	0.138	0.260	0.103	0.226	0.132	0.270	0.097	0.279	0.060
8	0.164	0.309	0.120	0.270	0.156	0.322	0.115	0.332	0.072
10	0.190	0.359	0.137	0.313	0.180	0.373	0.133	0.385	0.083
$\frac{1}{4}$	0.250	0.472	0.175	0.414	0.237	0.492	0.175	0.507	0.110
$\frac{5}{16}$	0.312	0.590	0.216	0.518	0.295	0.615	0.218	0.635	0.138
$\frac{3}{8}$	0.375	0.708	0.256	0.622	0.355	0.740	0.261	0.762	0.165
$\frac{1}{2}$	0.500	0.813	0.355	—	—	—	—	—	—

TABLE 8-1 (cont'd.)

Screw Number	Diameter	Flat Head 82°		Hex Head		Set Screw Hex Socket	
		D	H	F	H	F	C
0	0.060	0.119	0.035	—	—	0.028	0.033
1	0.073	0.146	0.043	—	—	0.035	0.040
2	0.086	0.172	0.051	—	—	0.035	0.047
4	0.112	0.225	0.067	—	—	0.050	0.061
6	0.138	0.279	0.083	—	—	0.062	0.074
8	0.164	0.332	0.100	—	—	0.078	0.087
10	0.190	0.385	0.116	—	—	0.094	0.102
$\frac{1}{4}$	0.250	0.507	0.153	0.438	0.163	0.125	0.132
$\frac{5}{16}$	0.312	0.635	0.191	0.500	0.211	0.156	0.172
$\frac{3}{8}$	0.375	0.762	0.230	0.562	0.243	0.188	0.212
$\frac{1}{2}$	0.500	0.875	0.223	0.750	0.323	0.250	0.291

D Diameter of head
H Height of head
F Distance across flats
C Diameter of cup

TABLE 8-2 Maximum sizes for nuts and washers

Screw Number	Hexagon Nuts		Split Lock Washer		Flat Washer		Int. Tooth Lock Washer		Ext. Tooth Lock Washer	
	F	T	D	T	D	T	D	T	D	T
0	.125	.050	.123	.022	.197	.025	——	——	——	——
1	.156	.050	.162	.020	——	——	——	——	——	——
2	.187	.066	.172	.020	.260	.025	.200	.015	——	——
4	.250	.098	.209	.025	.322	.040	.270	.019	.260	.019
6	.312	.114	.250	.031	.385	.065	.295	.021	.320	.022
8	.344	.130	.293	.040	.385	.065	.340	.023	.381	.023
10	.375	.130	.334	.047	.572	.080	.381	.025	.410	.025
$\frac{1}{4}$	.437	.193	.489	.062	.760	.080	.478	.028	.510	.028
$\frac{5}{16}$	.562	.225	.586	.078	.885	.104	.610	.034	.610	.034
$\frac{3}{8}$	.625	.257	.683	.094	1.010	.104	.692	.040	.694	.040
$\frac{1}{2}$	——	——	.873	.125	1.385	.132	.900	.045	.900	.045

T Thickness
F Distance Across Flats
D External Diameter

TABLE 8-3 Recommended screw sizes

Weight To Be Supported	4 Screws	2 Screws
Less than 1.5 lb	#4	#6
1.5–3	#6	#8
3–6.5	#8	#10
6.5–11	#10	$\frac{1}{4}$ in.
11–20	$\frac{1}{4}$ in.	——

TABLE 8-4 Wire color code

Black	0
Brown	1
Red	2
Orange	3
Yellow	4
Green	5
Blue	6
Violet	7
Gray	8
White	9

TABLE 8-5 Standard bare copper wire table (B&S)

Gauge (AWG) or (B&S)	Diameter, inches			Area	Weight	Length	Resistance at 68°F			Gauge (AWG) or (B&S)
	Min.	Nom.	Max.	Circular mils	Pounds per M′	Feet per lb.	Ohms per M′	Feet per ohm	Ohms per lb.	
0000	.4554	.4600	.4646	211600.	640.5	1.561	.04901	20400.	.00007652	0000
000	.4055	.4096	.4137	167800.	507.9	1.968	.06180	16180.	.0001217	000
00	.3612	.3648	.3684	133100.	402.8	2.482	.07793	12830.	.0001935	00
0	.3217	.3249	.3281	105500.	319.5	3.130	.09827	10180.	.0003076	0
1	.2864	.2893	.2922	83690.	253.3	3.947	.1239	8070.	.0004891	1
2	.2550	.2576	.2602	66370.	200.9	4.977	.1563	6400.	.0007778	2
3	.2271	.2294	.2317	52640.	159.3	6.276	.1970	5075.	.001237	3
4	.2023	.2043	.2063	41740.	126.4	7.914	.2485	4025.	.001966	4
5	.1801	.1819	.1837	33100.	100.2	9.980	.3133	3192.	.003127	5
6	.1604	.1620	.1636	26250.	79.46	12.58	.3951	2531.	.004972	6
7	.1429	.1443	.1457	20820.	63.02	15.87	.4982	2007.	.007905	7
8	.1272	.1285	.1298	16510.	49.98	20.01	.6282	1592.	.01257	8
9	.1133	.1144	.1155	13090.	39.63	25.23	.7921	1262.	.01999	9
10	.1009	.1019	.1029	10380.	31.43	31.82	.9989	1001.	.03178	10
11	.08983	.09074	.09165	8234.	24.92	40.12	1.260	794.	.05053	11
12	.08000	.08081	.08162	6530.	19.77	50.59	1.588	629.6	.08035	12
13	.07124	.07196	.07268	5178.	15.68	63.80	2.003	499.3	.1278	13
14	.06344	.06408	.06472	4107.	12.43	80.44	2.525	396.0	.2032	14
15	.05650	.05707	.05764	3257.	9.858	101.4	3.184	314.0	.3230	15
16	.05031	.05082	.05133	2583.	7.818	127.9	4.016	249.0	.5136	16
17	.04481	.04526	.04571	2048.	6.200	161.3	5.064	197.5	.8167	17
18	.03990	.04030	.04070	1624.	4.917	203.4	6.385	156.5	1.299	18
19	.03553	.03589	.03625	1288.	3.899	256.5	8.051	124.2	2.065	19
20	.03164	.03196	.03228	1022.	3.092	323.4	10.15	98.5	3.283	20

TABLE 8-5 (cont'd.)

21	.02818	.02846	.02874	810.1	2.452	407.8	12.80	78.11	5.221	21
22	.02510	.02535	.02560	642.4	1.945	514.2	16.14	61.95	8.301	22
23	.02234	.02257	.02280	509.5	1.542	648.4	20.36	49.13	13.20	23
24	.01990	.02010	.02030	404.0	1.223	817.7	25.67	38.96	20.99	24
25	.01770	.01790	.01810	320.4	.9699	1031.	32.37	30.90	33.37	25
26	.01578	.01594	.01610	254.1	.7692	1300.	40.81	24.50	53.06	26
27	.01406	.01420	.01434	201.5	.6100	1639.	51.47	19.43	84.37	27
28	.01251	.01264	.01277	159.8	.4837	2067.	64.90	15.41	134.2	28
29	.01115	.01126	.01137	126.7	.3836	2607.	81.83	12.22	213.3	29
30	.00993	.01003	.01013	100.5	.3042	3287.	103.2	9.691	339.2	30
31	.008828	.008928	.009028	79.7	.2413	4145.	130.1	7.685	539.3	31
32	.007850	.007950	.008050	63.21	.1913	5227.	164.1	6.095	857.6	32
33	.006980	.007080	.007180	50.13	.1517	6591.	206.9	4.833	1364.	33
34	.006205	.006305	.006405	39.75	.1203	8310.	260.9	3.833	2168.	34
35	.005515	.005615	.005715	31.52	.09542	10480.	329.0	3.040	3448.	35
36	.004900	.005000	.005100	25.00	.07568	13210.	414.8	2.411	5482.	36
37	.004353	.004453	.004553	19.83	.06001	16660.	523.1	1.912	8717.	37
38	.003865	.003965	.004065	15.72	.04759	21010.	659.6	1.516	13860.	38
39	.003431	.003531	.003631	12.47	.03774	26500.	831.8	1.202	22040.	39
40	.003045	.003145	.003245	9.888	.02993	33410.	1049.	0.9534	35040.	40
41	.00270	.00280	.00290	7.8400	.02373	42140.	1323.	.7559	55750.	41
42	.00239	.00249	.00259	6.2001	.01877	53270.	1673.	.5977	89120.	42
43	.00212	.00222	.00232	4.9284	.01492	67020.	2104.	.4753	141000.	43
44	.00187	.00197	.00207	3.8809	.01175	85100.	2672.	.3743	227380.	44
45	.00166	.00176	.00186	3.0976	.00938	106600.	3348.	.2987	356890.	45
46	.00147	.00157	.00167	2.4649	.00746	134040.	4207.	.2377	563900.	46

TABLE 8-6 Insulated copper wire table

AWG of Bare Copper Conductor	Diameter of Bare Copper Conductor		Insulation									
			Enamel & Single Cotton		Enamel & Single Silk		Enamel & Nylon		Resin & Single Cotton		Resin & Nylon	
			+Diameter over Insulation, inches									
	Min.	Max.	Min.	Max.	Min.	Max.	Min.	Max.	Min.	Max.	Min.	Max.
4	.2023	.2063	.2110	.2173	—	—	—	—	.2110	.2173	—	—
5	.1801	.1837	.1888	.1947	—	—	—	—	.1888	.1947	—	—
6	.1604	.1636	.1690	.1745	—	—	—	—	.1690	.1745	—	—
7	.1429	1457	.1514	.1565	—	—	—	—	.1514	.1565	—	—
8	.1272	.1298	.1356	.1404	*	*	*	*	.1356	.1404	—	—
9	.1133	.1155	.1209	.1251	*	*	*	*	.1208	.1251	—	—
10	.1009	.1029	.1075	.1114	*	*	*	*	.1075	.1114	—	—
11	.0898	.0916	.0960	.0996	*	*	*	*	.0956	.0991	—	—
12	.0800	.0816	.0861	.0895	*	*	*	*	.0857	.0890	—	—
13	.0713	.0727	.0774	.0805	*	*	*	*	.0770	.0800	—	—
14	.0635	.0647	.0696	.0725	*	*	*	*	.0692	.0720	—	—
15	.0565	.0577	.0625	.0654	.0590	.0619	.0591	.0621	.0621	.0649	.0596	.0628
16	.0503	.0513	.0562	.0589	.0527	.0554	.0528	.0556	.0558	.0584	.0533	.0563
17	.0448	.0458	.0507	.0533	.0472	.0498	.0473	.0500	.0503	.0528	.0478	.0507

TABLE 8-6 (cont'd.)

18	.0399	.0407	.0457	.0481	.0422	.0446	.0423	.0448	.0453	.0476	.0428	.0455
19	.0355	.0363	.0413	.0437	.0378	.0402	.0379	.0404	.0409	.0432	.0384	.0411
20	.0317	.0323	.0374	.0396	.0339	.0361	.0340	.0363	.0370	.0391	.0345	.0370
21	.0282	.0288	.0339	.0361	.0304	.0326	.0305	.0328	.0335	.0356	.0310	.0335
22	.0250	.0256	.0303	.0323	.0272	.0293	.0273	.0295	.0303	.0323	.0278	.0302
23	.0224	.0228	.0276	.0294	.0245	.0264	.0246	.0266	.0276	.0294	.0251	.0273
24	.0199	.0203	.0251	.0268	.0220	.0238	.0221	.0240	.0251	.0268	.0226	.0247
25	.0177	.0181	.0224	.0240	.0198	.0215	.0199	.0217	.0224	.0240	.0204	.0224
26	.0157	.0161	.0203	.0219	.0177	.0194	.0178	.0196	.0203	.0219	.0183	.0203
27	.0141	.0143	.0187	.0201	.0161	.0176	.0162	.0178	.0187	.0201	.0167	.0185
28	.0125	.0127	.0170	.0184	.0144	.0159	.0145	.0161	.0170	.0184	.0150	.0168
29	.0112	.0114	.0157	.0171	.0131	.0146	.0132	.0148	.0157	.0171	.0137	.0155
30	.0099	.0101	.0143	.0157	.0117	.0132	.0118	.0134	.0143	.0157	.0123	.0141
31	.0088	.0090	.0132	.0145	.0106	.0120	.0107	.0122	.0132	.0146	.0112	.0130
32	.0079	.0081	.0123	.0136	.0097	.0111	.0098	.0113	.0123	.0136	.0103	.0120
33	.0070	.0072	.0113	.0126	.0087	.0101	.0088	.0103	.0114	.0127	.0094	.0111
34	.0062	.0064	.0105	.0117	.0079	.0092	.0080	.0094	.0106	.0118	.0086	.0102
35	.0055	.0057	.0097	.0109	.0071	.0084	.0072	.0086	.0098	.0110	.0078	.0094
36	.0049	.0051	.0090	.0101	.0065	.0078	.0066	.0080	.0088	.0099	.0072	.0088
37	.0044	.0046	.0084	.0095	.0059	.0072	.0060	.0074	.0083	.0093	.0067	.0082
38	.0039	.0041	.0079	.0090	.0054	.0067	.0055	.0069	.0077	.0087	.0061	.0076
39	.0034	.0036	.0073	.0084	.0048	.0061	.0049	.0063	.0072	.0082	.0056	.0071
40	.0030	.0032	.0069	.0080	.0044	.0057	.0045	.0059	.0068	.0078	.0052	.0067

*Special construction—not standardized.

TABLE 8-7 Stranded equivalents of solid wires

(Area in circular mils)

Solid Wire		Stranded
AWG No.	Area	
12	6530	41/28, 65/30
14	4107	26/28, 41/30, 65/32
16	2583	7/24, 10/26, 16/28
18	1624	7/26, 10/28, 16/30
20	1022	7/28, 10/30, 26/34
22	642.4	7/30, 10/32, 16/34
24	404.0	7/32, 16/36, 41/40
26	254.1	7/34, 10/36
28	159.8	7/36, 16/40
30	100.5	7/38
32	63.21	7/40

TABLE 8-8 Current-carrying capacity

Wire Size	Current
#10	35 amperes
12	26
14	20
16	15
18	11
20	7
22	5
24	3
26	2
28	1

TABLE 9-1 Laminates for PWBs

Designation NEMA	Designation MIL	Resin	Filler Material	Comments
XXXP		Phenolic	Paper	
XXXPC		Phenolic	Paper	May be hot punched
FR-2		Phenolic	Paper	Flame retardant (FR)
FR-3	PX	Epoxy	Paper	FR, cold punched
FR-3	PH	Epoxy	Paper	FR, hot punched
FR-4	GF	Epoxy	Glass fabric	FR
FR-5	GH	Epoxy	Glass fabric	FR, temp resistant
G-10	GE	Epoxy	Glass fabric	
G-11	GB	Epoxy	Glass fabric	Temp resistant
	GC	Polyester	Glass fabric and fiber	FR
	GP	PTFE*	Glass fiber	FR
	GT	PTFE*	Glass fabric	FR

*PTFE: polytetrafluoroethylene. *Teflon* is Dupont's trade name for this material.

TABLE 9-2 Thickness tolerances

(All tolerances are in ± the indicated fraction of an inch.)

Overall Thickness	Paper: 1-oz Copper on Side	Paper: All Other	Glass
$\frac{1}{32}$ (0.031)	0.0040	0.0045	0.0065
$\frac{1}{16}$ (0.062)	0.0055	0.0060	0.0075
$\frac{3}{32}$ (0.093)	0.0070	0.0075	0.0090
$\frac{1}{8}$ (0.125)	0.0085	0.0090	0.0120
$\frac{1}{4}$ (0.250)	0.0120	0.0120	0.0220

TABLE 9-3 Thickness of copper cladding

Weight	Thickness
0.5 oz/sq ft	0.0007 in.
1	0.0014
2	0.0027
3	0.0040
5	0.0068
7	0.0094

TABLE 9-4 Conductor spacing

Altitudes	Voltage Difference	Minimum Spacing
Sea level to 10,000 feet (uncoated boards)	0–150 volts	0.025 in.
	151–300	0.050
	301–500	0.100
	over 500	0.0002 in./volt
Over 10,000 feet (uncoated boards)	0–50	0.025
	51–100	0.060
	101–170	0.125
	171–250	0.250
	251–500	0.500
	over 500	0.001 in./volt
All altitudes (coated boards)	0–30	0.010
	31–50	0.015
	51–150	0.020
	151–300	0.030
	301–500	0.060
	over 500	0.00012 in./volt

TABLE 9-5 Tolerances

Dimension	Tolerance
Diameter of drilled hole	±0.003 in.
Diameter of punched hole	±0.004
Diameter of hole after plating	±0.003
Hole-to-hole distance	±0.010
Hole-to-pattern distance	±0.015
Front-to-back pattern registration	±0.020
Conductor to outside edge	±0.015
Reference hole to edge	±0.015
Outside dimensions	±0.015
Line width and spacing	±0.015

TABLE 9-6 Recommended mounting hole spacing

Component	Body Length	Recommended Spacing
0.1 watt resistor	0.140 in.	0.500 in.
0.25 watt resistor	0.375	0.700
1 watt resistor	0.562	0.900
Diode	0.265	0.600
Tantalum capacitor	1.188	1.500
Tantalum capacitor	1.375	1.700
Tantalum capacitor	1.937	2.300
Paper capacitor	0.843	1.200
Paper capacitor	1.000	1.300
Paper capacitor	1.375	1.700

TABLE 10-1 Fixed resistor selection guide

Type	Power and Max Voltage Ratings	Ohmic Range	Temperature Range	Lead Configuration	Max Body Size
Composition (Insulated)	$\frac{1}{8}$ w/150 v	10 to 22 M	70°–130°	Axial lead	.160 × .066
(MIL-R-11)	$\frac{1}{4}$ w/250 v	2.7 to 22 M	70°–150°	Axial lead	.281 × .098
	$\frac{1}{2}$ w/350 v	2.7 to 22 M	70°–130°	Axial lead	.416 × .161
	1 w/500 v	2.7 to 22 M	70°–130°	Axial lead	.593 × .240
	2 w/500 v	10 to 22 M	70°–130°	Axial lead	.728 × .336
Film (High Stability)	$\frac{1}{20}$ w/200 v	10 to .100 M	125°–175°	Axial lead	.160 × .070
(MIL-R-10509)		10 to .301 M	70°–165°; 125°–175°	Axial lead	.281 × .140
	See	10 to 1 M	70°–165°; 125°–175°	Axial lead	.437 × .165
	Section	10 to 2 M	70°–165°; 125°–175°	Axial lead	.656 × .249
	102	10 to 2.49 M	70°–165°; 125°–175°	Axial lead	.875 × .328
		10 to 5.11 M	70°–150°; 125°–175°	Axial lead	1.124 × .437
Film (Power Type)	2 w/350 v	10 to .178 M	25°–275°	Axial lead	.656 × .235
(MIL-R-11804)	4 w/500 v	10 to .464 M	25°–275°	Axial lead	1.000 × .359
	8 w/750 v	21.5 to 1.0 M	25°–275°	Axial lead	2.124 × .359
Film, Insulated	$\frac{1}{4}$ w/250 v	47 to .15 M	70°–150°	Axial lead	.281 × .098
(MIL-R-22684)	$\frac{1}{2}$ w/350 v	4.7 to .47 M	70°–150°	Axial lead	.416 × .161
	1 w/500 v	10 to 1.0 M	70°–150°	Axial lead	.593 × .205
	2 w/500 v	10 to 1.5 M	70°–150°	Axial lead	.728 × .336
Composition (Insulated),	$\frac{1}{4}$ w/250 v	10 to 1 M	70°–130°	Axial lead	.281 × .098
Established Reliability	$\frac{1}{2}$ w/350 v	10 to 1 M	70°–130°	Axial lead	.415 × .161
(MIL-R-39008)	1 w/500 v	10 to 1 M	70°–130°	Axial lead	.593 × .240
Film,	$\frac{1}{20}$ w/200 v	49.9 to .100 M	125°–175°	Axial lead	.170 × .080
Established Reliability	$\frac{1}{10}$ w/200 v	10 to .301 M	70°–165°; 125°–175°	Axial lead	.281 × .140
(MIL-R-55182)	$\frac{1}{8}$ w/200 v	10 to .200 M	70°–165°; 125°–175°	Axial lead	.343 × .170
	$\frac{1}{8}$ w/250 v	10 to 1.0 M	70°–165°; 125°–175°	Axial lead	.437 × .165
	$\frac{1}{4}$ w/300 v	10 to 2.0 M	70°–165°; 125°–175°	Axial lead	.656 × .250
	$\frac{1}{2}$ w/350 v	10 to 2.49 M	70°–165°; 125°–175°	Axial lead	.875 × .328
Film (Insulated),	$\frac{1}{4}$ w/250 v	47 to .15 M	70°–150°	Axial lead	.281 × .098
Established Reliability	$\frac{1}{2}$ w/350 v	10 to .47 M	70°–150°	Axial lead	.416 × .161
(MIL-R-39017)	1 w/500 v	10 to 1.0 M	70°–150°	Axial lead	.593 × .205
Wirewound (Accurate)	$\frac{1}{2}$ w/600 v	.1 to 1.5 M	125°–145°	Axial lead	1.062 × .406
(MIL-R-93)	$\frac{1}{3}$ w/300 v	.1 to .750 M	125°–145°	Axial lead	.812 × .418
	.15 w/—	.1 to .226 M	125°–145°	Axial lead	.562 × .281

	$\frac{1}{8}$ w/—	.1 to .100 M	125°–145°	PC pins	.343 × .281
Wirewound (Power Type) (MIL-R-26)	11 w	.10 to 5.6 K	25°–350°	Radial tab	1.812 × .500
	14 w	.10 to 6.8 K	25°–350°	Radial tab	1.562 × .594
	26 w	.10 to 18 K	25°–350°	Radial tab	3.062 × .594
	55 w	.10 to 39 K	25°–350°	Radial tab	4.062 × .906
	113 w	.10 to 82 K	25°–350°	Radial tab	6.062 × 1.312
	159 w	.10 to .15 M	25°–350°	Radial tab	8.062 × 1.312
	210 w	.10 to .18 M	25°–350°	Radial tab	10.562 × 1.312
	7 w	.10 to 8.2 K	25°–350°	Axial lead	1.469 × .562
	14 w	.10 to 15 K	25°–350°	Axial lead	2.094 × .562
	6.5 w	.10 to 3.3 K	25°–350°	Axial lead	1.094 × .374
	11 w	.10 to 8.2 K	25°–350°	Axial lead	1.937 × .437
	3 w	.10 to 820 Ω	25°–350°	Axial lead	.562 × .249
	1.5 w	.10 to 3.16 K	25°–350°	Axial lead	.437 × .141
	6.5 w	.10 to 38.3 K	25°–350°	Axial lead	1.094 × .374
	11 w	.10 to 90.9 K	25°–350°	Axial lead	1.937 × .437
	3 w	.10 to 10.5 K	25°–350°	Axial lead	.562 × .249
Wirewound (Power Type, Chassis Mounted) (MIL-R-18546)	5	.10 to 3.32 K	25°–275°	Axial lug	.662 × .677 × .351
	10 Mtd on	.10 to 5.62 K	25°–275°	Axial lug	.812 × .843 × .437
	20 metal	.10 to 12.1 K	25°–275°	Axial lug	1.124 × 1.125 × .593
	30 chassis	.10 to 38.3 K	25°–275°	Axial lug	2.000 × 1.187 × .656
	75	.10 to 34.8 K	25°–275°	Axial screw thread	3.531 × 2.843 × 1.761
Wirewound (Accurate), Established Reliability (MIL-R-39005)	.5 w/600 v	.1 to 1.2 M	125°–145°	Axial lead	1.062 × .406
	$\frac{1}{3}$ w/300 v	.1 to 1.75 M	125°–145°	Axial lead	.812 × .406
	$\frac{1}{4}$ w/300 v	.1 to .850 M	125°–145°	Axial lead	.812 × .281
	.15 w/200 v	.1 to .525 M	125°–145°	Axial lead	.562 × .281
	$\frac{1}{8}$ w/150 v	.1 to .350 M	125°–145°	Axial lead	.406 × .281
	$\frac{1}{8}$ w/150 v	.1 to .270 M	125°–145°	PC pins	.344 × .281
Wirewound (Power Type), Established Reliability (MIL-R-39007)	5.0 w	1.0 to 12.1 K	25°–275°	Axial lead	1.094 × .374
	10 w	1.0 to 38.3 K	25°–275°	Axial lead	1.937 × .438
	2.5 w	1.0 to 3.48 K	25°–275°	Axial lead	.562 × .250
	1 w	1.0 to 1.21 K	25°–275°	Axial lead	.437 × .124
	2 w	1.0 to 6.04 K	25°–275°	Axial lead	.874 × .218
Wirewound (Power Type, Chassis Mounted), Established Reliability (MIL-R-39009)	10 w Mtd on	.1 to 5.62 K	25°–275°	Axial lug	.812 × .843 × .437
	15 w metal	.1 to 12.1 K	25°–275°	Axial lug	1.124 × 1.124 × .593
	30 w chassis	.1 to 38.3 K	25°–275°	Axial lug	2.000 × 1.187 × .656

TABLE 10-2 Variable resistor selection guide

Type	Power Rating (watts)	Nominal Total Resistance	Temperature Range	Mounting Data	Maximum Body Size (Dia. × Depth) (inches)
Composition (MIL-R-94)	2, 1	50 to 5 M	70°–120°C	Shaft and panel seal and locking bush.	1.156 × .750
	$\frac{1}{2}$, $\frac{1}{4}$	100 to 5 M		Standard and locking bush.	.515 × .468
Wirewound (Low Operating Temperature) (MIL-R-19)	2.0, 1.1	3 to 15 K	40°–105°C	Shaft and panel seal and locking bush.	1.313 × .703
	4.0, 2.2	3 to 25 K			1.719 × .813
Wirewound (Power Type) (MIL-R-22)	6.25	1.0 to 3.5 K	25°–340°C	Std bush. & lkg bush.	.906 × .751
	25	2.0 to 5 K			1.680 × 1.410
	50	1.0 to 10 K			2.410 × 1.440
	75	2.0 to 10 K			2.810 × 1.780
	100	2.0 to 10 K			3.190 × 1.780
	150	2.0 to 10 K			4.060 × 2.030
Wirewound, Semiprecision (MIL-R-39002)	1.5	10 to 50 K	85°–135°C	Std bush. & lkg bush.	.570 × .650
Wirewound (Lead Screw Actuated) (MIL-R-27208)	$\frac{3}{4}$	10 to 10 K	85°–150°C	PC pin; mtg hole	1.260 × .200 × .330
	$\frac{3}{4}$	50 to 10 K			.510 × .197 × .510
Nonwirewound (Lead Screw Actuated) (MIL-R-22097)	$\frac{1}{4}$, $\frac{3}{4}$	100 to 1 M		PC pin; mtg hole	1.282 × .300 × .374
	$\frac{3}{4}$	100 to 1 M			.510 × .197 × .510
Wirewound (Lead Screw Actuated), Established Reliability (MIL-R-39015)	$\frac{3}{4}$	100 to 20 K	85°–150°C	PC pin; mtg hole	1.260 × .200 × .330
	$\frac{3}{4}$	100 to 20 K			.510 × .197 × .510

TABLE 11-1 Rigid rectangular waveguides and flanges

EIA WG Designation WR( )	Material Alloy	JAN WG Designation RG( )/U	JAN Flange Designation: Choke UG( )/U	JAN Flange Designation: Cover UG( )/U	Dimensions (inches): Inside	Tolerance	Outside	Tolerance	Wall Thickness Nominal
2300	Alum.				23.000–11.500	±0.020	23.250–11.750	±0.020	0.125
2100	Alum.				21.000–10.500	±0.020	21.250–10.750	±0.020	0.125
1800	Alum.	201			18.000–9.000	±0.020	18.250–9.250	±0.020	0.125
1500	Alum.	202			15.000–7.500	±0.015	15.250–7.750	±0.015	0.125
1150	Alum.	203			11.500–5.750	±0.015	11.750–6.000	±0.015	0.125
975	Alum.	204			9.750–4.875	±0.010	10.000–5.125	±0.010	0.125
770	Alum.	205			7.700–3.850	±0.005	7.950–4.100	±0.005	0.125
650	Brass Alum.	69 103		417A 418A	6.500–3.250	±0.005	6.660–3.410	±0.005	0.080
510					5.100–2.550	±0.005	5.260–2.710	±0.005	0.080
430	Brass Alum.	104 105		435A 437A	4.300–2.150	±0.005	4.460–2.310	±0.005	0.080
340	Brass Alum.	112 113		553 554	3.400–1.700	±0.005	3.560–1.860	±0.005	0.080
284	Brass Alum.	48 75	54A 585	53 584	2.840–1.340	±0.005	3.000–1.500	±0.005	0.080
229					2.290–1.145	±0.005	2.418–1.273	±0.005	0.064
187	Brass Alum.	49 95	148B 406A	149A 407	1.872–0.872	±0.005	2.000–1.000	±0.005	0.064
159					1.590–0.795	±0.004	1.718–0.923	±0.004	0.064
137	Brass Alum.	50 106	343A 440A	344 441	1.372–0.622	±0.004	1.500–0.750	±0.004	0.064

TABLE 11-1 (cont'd.)

EIA WG Designation WR( )	Material Alloy	JAN WG Designation RG( )/U	JAN Flange Designation: Choke UG( )/U	JAN Flange Designation: Cover UG( )/U	Dimensions (inches): Inside	Tol.	Outside	Tol.	Wall Thickness Nominal
112	Brass	51	52A	51	1.122–0.497	±0.004	1.250–0.625	±0.004	0.064
	Alum.	68	137A	138					
90	Brass	52	40A	39	0.900–0.400	±0.003	1.000–0.500	±0.003	0.050
	Alum.	67	136A	135					
75					0.750–0.375	±0.003	0.850–0.475	±0.003	0.050
	Brass	91	541	419					
62	Alum.	—	—	—	0.622–0.311	±0.0025	0.702–0.391	±0.003	0.040
	Silver	107	—	—					
51					0.510–0.255	±0.0025	0.590–0.335	±0.003	0.040
	Brass	53	596	595					
42	Alum.	121	598	597	0.420–0.170	±0.0020	0.500–0.250	±0.003	0.040
	Silver	66	—	—					
34					0.340–0.170	±0.0020	0.420–0.250	±0.003	0.040
	Brass	—	600	599					
28	Alum.	—	—	—	0.280–0.140	±0.0015	0.360–0.220	±0.002	0.040
	Silver	96	—	—					
22	Brass	—		383	0.224–0.112	±0.0010	0.304–0.192	±0.002	0.040
	Silver	97		—					
19					0.188–0.094	±0.0010	0.268–0.174	±0.002	0.040
15	Brass	—		385	0.148–0.074	±0.0010	0.228–0.154	±0.002	0.040
	Silver	93		—					
12	Brass	—		387	0.122–0.061	±0.0005	0.202–0.141	±0.002	0.040
	Silver	99		—					
10					0.100–0.050	±0.0005	0.180–0.130	±0.002	0.040
8	Silver	138	—	—	0.080–0.040	±0.0003	0.156 DLA	±0.001	—
7	Silver	136	—	—	0.065–0.0325	±0.00025	0.156 DLA	±0.001	—
5	Silver	135	—	—	0.051–0.0255	±0.00025	0.156 DLA	±0.001	—
4	Silver	137	—	—	0.043–0.0215	±0.00020	0.156 DLA	±0.001	—
3	Silver	139	—	—	0.034–0.0170	±0.00020	0.156 DLA	±0.001	—

TABLE 11-2 Range in guide wavelength

Waveguide	Frequency	Range of λ_g
WR770	1 GHz	18.360–18.385 inches
WR284	3	5.445–5.463
WR187	5	3.036–3.046
WR90	9	1.907–1.922
WR62	16	0.914–0.918

TABLE 11-3 Dielectrics for stripline

Material	ϵ
Teflon	2.08
Polyolefin	2.32
Rexolite	2.65
Textolite	2.56

APPENDIX B

Other Useful Tables

Temperature conversion table

−60 to 4			5 to 53			54 to 120			130 to 600		
C.		F.	C.		F.	C.		F.	C.		F.
−51.1	**−60**	−76.0	−15.0	**5**	41.0	12.2	**54**	129.2	54	**130**	266
−48.2	**−55**	−67.0	−14.4	**6**	42.8	12.8	**55**	131.0	60	**140**	284
−45.5	**−50**	−58.0	−13.9	**7**	44.6	13.3	**56**	132.8	66	**150**	302
−42.7	**−45**	−49.0	−13.3	**8**	46.4	13.9	**57**	134.6	71	**160**	320
−40.0	**−40**	−40.0	−12.8	**9**	48.6	14.4	**58**	136.4	77	**170**	338
−39.4	**−39**	−38.2	−12.2	**10**	50.0	15.0	**59**	138.2	82	**180**	356
−38.8	**−38**	−36.4	−11.7	**11**	51.8	15.6	**60**	140.0	88	**190**	374
−38.2	**−37**	−34.6	−11.1	**12**	53.6	16.1	**61**	141.8	93	**200**	392
−37.7	**−36**	−32.8	−10.6	**13**	55.4	16.7	**62**	143.6	99	**210**	410
−37.1	**−35**	−31.0	−10.0	**14**	57.2	17.2	**63**	145.4	100	**212**	413
−36.6	**−34**	−29.2	−9.44	**15**	59.0	17.8	**64**	147.2	104	**220**	428
−36.0	**−33**	−27.4	−8.89	**16**	60.8	18.3	**65**	149.0	110	**230**	446
−35.5	**−32**	−25.6	−8.33	**17**	62.6	18.9	**66**	150.8	116	**240**	464
−34.9	**−31**	−23.8	−7.78	**18**	64.4	19.4	**67**	152.6	121	**250**	482
−34.4	**−30**	−22.0	−7.22	**19**	66.2	20.0	**68**	154.4	127	**260**	500
−33.8	**−29**	−20.2	−6.67	**20**	68.0	20.6	**69**	156.2	132	**270**	518
−33.2	**−28**	−18.4	−6.11	**21**	69.8	21.1	**70**	158.0	138	**280**	536
−32.6	**−27**	−16.6	−5.56	**22**	71.6	21.7	**71**	159.8	143	**290**	554
−32.1	**−26**	−14.8	−5.00	**23**	73.4	22.2	**72**	161.6	149	**300**	572
−31.5	**−25**	−13.0	−4.44	**24**	75.2	22.8	**73**	163.4	154	**310**	590
−31.0	**−24**	−11.2	−3.89	**25**	77.0	23.3	**74**	165.2	160	**320**	608
−30.4	**−23**	− 9.4	−3.33	**26**	78.8	23.9	**75**	167.0	166	**330**	626
−29.9	**−22**	− 7.6	−2.78	**27**	80.6	24.4	**76**	168.8	171	**340**	644
−29.3	**−21**	− 5.8	−2.22	**28**	82.4	25.0	**77**	170.6	177	**350**	662
−28.8	**−20**	− 4.0	−1.67	**29**	84.2	25.6	**78**	172.4	182	**360**	680
−28.3	**−19**	− 2.2	−1.11	**30**	86.0	26.1	**79**	174.2	188	**370**	698
−27.7	**−18**	−0.40	−0.56	**31**	87.8	26.7	**80**	176.0	193	**380**	716
−27.1	**−17**	1.4	0	**32**	89.6	27.2	**81**	177.8	199	**390**	734
−26.6	**−16**	3.2	0.56	**33**	91.4	27.8	**82**	179.6	204	**400**	752
−26.0	**−15**	5.0	1.11	**34**	93.2	28.3	**83**	181.4	210	**410**	770

Temperature conversion table

−25.5	**−14**	6.8	1.67	**35**	95.0	28.9	**84**	183.2	216	**420**	788
−24.9	**−13**	8.6	2.22	**36**	96.8	29.4	**85**	185.0	221	**430**	806
−24.4	**−12**	10.4	2.78	**37**	98.6	30.0	**86**	186.8	227	**440**	824
−23.8	**−11**	12.2	3.33	**38**	100.4	30.6	**87**	188.6	232	**450**	842
−23.3	**−10**	14.0	3.89	**39**	102.2	31.1	**88**	190.4	238	**460**	860
−22.6	**− 9**	15.8	4.44	**40**	104.0	31.7	**89**	192.2	243	**470**	878
−22.1	**− 8**	17.6	5.00	**41**	105.8	32.2	**90**	194.0	249	**480**	896
−21.5	**− 7**	19.4	5.56	**42**	107.6	32.8	**91**	195.8	254	**490**	914
−21.0	**− 6**	21.2	6.11	**43**	109.4	33.3	**92**	197.6	260	**500**	932
−20.4	**− 5**	23.0	6.67	**44**	111.2	33.9	**93**	199.4	266	**510**	950
−19.9	**− 4**	24.8	7.22	**45**	113.0	34.4	**94**	201.2	271	**520**	968
−19.3	**− 3**	26.6	7.78	**46**	114.8	35.0	**95**	203.0	277	**530**	986
−18.8	**− 2**	28.4	8.33	**47**	116.6	35.6	**96**	204.8	282	**540**	1004
−18.3	**− 1**	30.2	8.89	**48**	118.4	36.1	**97**	206.6	288	**550**	1022
−17.8	**0**	32.0	9.44	**49**	120.2	36.7	**98**	208.4	293	**560**	1040
−17.2	**1**	33.8	10.0	**50**	122.0	37.2	**99**	210.2	299	**570**	1058
−16.7	**2**	35.6	10.6	**51**	123.8	37.8	**100**	212.0	304	**580**	1076
−16.1	**3**	37.4	11.1	**52**	125.6	43	**110**	230	310	**590**	1094
−15.6	**4**	39.2	11.7	**53**	127.4	49	**120**	248	316	**600**	1112

NOTE: The numbers in boldface type refer to the temperature either in degrees Centrigrade or Fahrenheit which it is desired to convert into the other scale. If converting from Fahrenheit degrees to Centrigrade degrees, the equivalent temperature will be found in the left column, while if converting from degrees Centrigrade to degrees Fahrenheit, the answer will be found in the column on the right.

INTERPOLATION FACTORS

C.		F.	C.		F.
0.56	**1**	1.8	3.33	**6**	10.8
1.11	**2**	3.6	3.89	**7**	12.6
1.67	**3**	5.4	4.44	**8**	14.4
2.22	**4**	7.2	5.00	**9**	16.2
2.78	**5**	9.0	5.56	**10**	18.0

Table of four-place logarithms of numbers

N	0	1	2	3	4	5	6	7	8	9
10	0000	0043	0086	0128	0170	0212	0253	0294	0334	0374
11	0414	0453	0492	0531	0569	0607	0645	0682	0719	0755
12	0792	0828	0864	0899	0934	0969	1004	1038	1072	1106
13	1139	1173	1206	1239	1271	1303	1335	1367	1399	1430
14	1461	1492	1523	1553	1584	1614	1644	1673	1703	1732
15	1761	1790	1818	1847	1875	1903	1931	1959	1987	2014
16	2041	2068	2095	2122	2148	2175	2201	2227	2253	2279
17	2304	2330	2355	2380	2405	2430	2455	2480	2504	2529
18	2553	2577	2601	2625	2648	2672	2695	2718	2742	2765
19	2788	2810	2833	2856	2878	2900	2923	2945	2967	2989
20	3010	3032	3054	3075	3096	3118	3139	3160	3181	3201
21	3222	3243	3263	3284	3304	3324	3345	3365	3385	3404
22	3424	3444	3464	3483	3502	3522	3541	3560	3579	3598
23	3617	3636	3655	3674	3692	3711	3729	3747	3766	3784
24	3802	3820	3838	3856	3874	3892	3909	3927	3945	3962
25	3979	3997	4014	4031	4048	4065	4082	4099	4116	4133
26	4150	4166	4183	4200	4216	4232	4249	4265	4281	4298
27	4314	4330	4346	4362	4378	4393	4409	4425	4440	4456
28	4472	4487	4502	4518	4533	4548	4564	4579	4594	4609
29	4624	4639	4654	4669	4683	4698	4713	4728	4742	4757
30	4771	4786	4800	4814	4829	4843	4857	4871	4886	4900
31	4914	4928	4942	4955	4969	4983	4997	5011	5024	5038
32	5051	5065	5079	5092	5105	5119	5132	5145	5159	5172
33	5185	5198	5211	5224	5237	5250	5263	5276	5289	5302
34	5315	5328	5340	5353	5366	5378	5391	5403	5416	5428
35	5441	5453	5465	5478	5490	5502	5514	5527	5539	5551
36	5563	5575	5587	5599	5611	5623	5635	5647	5658	5670
37	5682	5694	5705	5717	5729	5740	5752	5763	5775	5786
38	5798	5809	5821	5832	5843	5855	5866	5877	5888	5899
39	5911	5922	5933	5944	5955	5966	5977	5988	5999	6010
40	6021	6031	6042	6053	6064	6075	6085	6096	6107	6117
41	6128	6138	6149	6160	6170	6180	6191	6201	6212	6222
42	6232	6243	6253	6263	6274	6284	6294	6304	6314	6325
43	6335	6345	6355	6365	6375	6385	6395	6405	6415	6425
44	6435	6444	6454	6464	6474	6484	6493	6503	6513	6522
45	6532	6542	6551	6561	6571	6580	6590	6599	6609	6618
46	6628	6637	6646	6656	6665	6675	6684	6693	6702	6712
47	6721	6730	6739	6749	6758	6767	6776	6785	6794	6803
48	6812	6821	6830	6839	6848	6857	6866	6875	6884	6893
49	6902	6911	6920	6928	6937	6946	6955	6964	6972	6981
50	6990	6998	7007	7016	7024	7033	7042	7050	7059	7067
51	7076	7084	7093	7101	7110	7118	7126	7135	7143	7152
52	7160	7168	7177	7185	7193	7202	7210	7218	7226	7235
53	7243	7251	7259	7267	7275	7284	7292	7300	7308	7316
54	7324	7332	7340	7348	7356	7364	7372	7380	7388	7396

Table of four-place logarithms of numbers (cont'd.)

N	0	1	2	3	4	5	6	7	8	9
55	7404	7412	7419	7427	7435	7443	7451	7459	7466	7474
56	7482	7490	7497	7505	7513	7520	7528	7536	7543	7551
57	7559	7566	7574	7582	7589	7597	7604	7612	7619	7627
58	7634	7642	7649	7657	7664	7672	7679	7686	7694	7701
59	7709	7716	7723	7731	7738	7745	7752	7760	7767	7774
60	7782	7789	7796	7803	7810	7818	7825	7832	7839	7846
61	7853	7860	7868	7875	7882	7889	7896	7903	7910	7917
62	7924	7931	7938	7945	7952	7959	7966	7973	7980	7987
63	7993	8000	8007	8014	8021	8028	8035	8041	8048	8055
64	8062	8069	8075	8082	8089	8096	8102	8109	8116	8122
65	8129	8136	8142	8149	8156	8162	8169	8176	8182	8189
66	8195	8202	8209	8215	8222	8228	8235	8241	8248	8254
67	8261	8267	8274	8280	8287	8293	8299	8306	8312	8319
68	8325	8331	8338	8344	8351	8357	8363	8370	8376	8382
69	8388	8395	8401	8407	8414	8420	8426	8432	8439	8445
70	8451	8457	8463	8470	8476	8482	8488	8494	8500	8506
71	8513	8519	8525	8531	8537	8543	8549	8555	8561	8567
72	8573	8579	8585	8591	8597	8603	8609	8615	8621	8627
73	8633	8639	8645	8651	8657	8663	8669	8675	8681	8686
74	8692	8698	8704	8710	8716	8722	8727	8733	8739	8745
75	8751	8756	8762	8768	8774	8779	8785	8791	8797	8802
76	8808	8814	8820	8825	8831	8837	8842	8848	8854	8859
77	8865	8871	8876	8882	8887	8893	8899	8904	8910	8915
78	8921	8927	8932	8938	8943	8949	8954	8960	8965	8971
79	8976	8982	8987	8993	8998	9004	9009	9015	9020	9025
80	9031	9036	9042	9047	9053	9058	9063	9069	9074	9079
81	9085	9090	9096	9101	9106	9112	9117	9122	9128	9133
82	9138	9143	9149	9154	9159	9165	9170	9175	9180	9186
83	9191	9196	9201	9206	9212	9217	9222	9227	9232	9238
84	9243	9248	9253	9258	9263	9269	9274	9279	9284	9289
85	9294	9299	9304	9309	9315	9320	9325	9330	9335	9340
86	9345	9350	9355	9360	9365	9370	9375	9380	9385	9390
87	9395	9400	9405	9410	9415	9420	9425	9430	9435	9440
88	9445	9450	9455	9460	9465	9469	9474	9479	9484	9489
89	9494	9499	9504	9509	9513	9518	9523	9528	9533	9538
90	9542	9547	9552	9557	9562	9566	9571	9576	9581	9586
91	9590	9595	9600	9605	9609	9614	9619	9624	9628	9633
92	9638	9643	9647	9652	9657	9661	9666	9671	9675	9680
93	9685	9689	9694	9699	9703	9708	9713	9717	9722	9727
94	9731	9736	9741	9745	9750	9754	9759	9763	9768	9773
95	9777	9782	9786	9791	9795	9800	9805	9809	9814	9818
96	9823	9827	9832	9836	9841	9845	9850	9854	9859	9863
97	9868	9872	9877	9881	9886	9890	9894	9899	9903	9908
98	9912	9917	9921	9926	9930	9934	9939	9943	9948	9952
99	9956	9961	9965	9969	9974	9978	9983	9987	9991	9996

Decibel conversion chart

Power Ratio	Voltage and Current Ratio	Decibels (+) (−)	Voltage and Current Ratio	Power Ratio
1.000	1.000	0	1.000	1.000
1.023	1.012	0.1	.9886	.9772
1.047	1.023	0.2	.9772	.9550
1.072	1.035	0.3	.9661	.9333
1.096	1.047	0.4	.9550	.9120
1.122	1.059	0.5	.9441	.8913
1.148	1.072	0.6	.9333	.8710
1.175	1.084	0.7	.9226	.8511
1.202	1.096	0.8	.9120	.8318
1.230	1.109	0.9	.9016	.8128
1.259	1.122	1.0	.8913	.7943
1.585	1.259	2.0	.7943	.6310
1.995	1.413	3.0	.7079	.5012
2.512	1.585	4.0	.6310	.3981
3.162	1.778	5.0	.5623	.3162
3.981	1.995	6.0	.5012	.2512
5.012	2.239	7.0	.4467	.1995
6.310	2.512	8.0	.3981	.1585
7.943	2.818	9.0	.3548	.1259
10.00	3.162	10.0	.3162	.1000
12.59	3.548	11.0	.2818	.07943
15.85	3.981	12.0	.2515	.06310
19.95	4.467	13.0	.2293	.05012
25.12	5.012	14.0	.1995	.03981
31.62	5.632	15.0	.1778	.03162
39.81	6.310	16.0	.1585	.02512
50.12	7.079	17.0	.1413	.01995
63.10	7.943	18.0	.1259	.01585
79.43	8.913	19.0	.1122	.01259
100.00	10.00	20.0	.1000	.01000
10^3	31.62	30.0	.03162	.00100
10^4	10^2	40.0	10^{-2}	10^{-4}
10^5	316.23	50.0	3.162×10^{-3}	10^{-5}
10^6	10^3	60.0	10^{-3}	10^{-6}
10^7	3.162×10^3	70.0	3.162×10^{-4}	10^{-7}
10^8	10^4	80.0	10^{-4}	10^{-8}
10^9	3.162×10^4	90.0	3.162×10^{-5}	10^{-9}
10^{10}	10^5	100.00	10^{-5}	10^{-10}

Table of natural trigonometric functions

DEGREES	Sin	Cos	Tan	Cot	
0°00′	.0000	1.0000	.0000	——	**90°00′**
10	029	000	029	343.8	50
20	058	000	058	171.9	40
30	.0087	1.0000	.0087	114.6	30
40	116	.9999	116	85.94	20
50	145	999	145	68.75	10
1°00′	.0175	.9998	.0175	57.29	**89°00′**
10	204	998	204	49.10	50
20	233	997	233	42.96	40
30	.0262	.9997	.0262	38.19	30
40	291	996	291	34.37	20
50	320	995	320	31.24	10
2°00′	.0349	.9994	.0349	28.64	**88°00′**
10	378	993	378	26.43	50
20	407	992	407	24.54	40
30	.0436	.9990	.0437	22.90	30
40	465	989	466	21.47	20
50	494	988	495	20.21	10
3°00′	.0523	.9986	.0524	19.08	**87°00′**
10	552	985	553	18.07	50
20	581	983	582	17.17	40
30	.0610	.9981	.0612	16.35	30
40	640	980	641	15.60	20
50	669	978	670	14.92	10
4°00′	.0698	.9976	.0699	14.30	**86°00′**
10	727	974	729	13.73	50
20	756	971	758	13.20	40
30	.0785	.9969	.0787	12.71	30
40	814	967	816	12.25	20
50	843	964	846	11.83	10
5°00′	.0872	.9962	.0875	11.43	**85°00′**
10	901	959	904	11.06	50
20	929	957	934	10.71	40
30	.0958	.9954	.0963	10.39	30
40	987	951	992	10.08	20
50	.1016	948	.1022	9.788	10
6°00′	.1045	.9945	.1051	9.514	**84°00′**
10	074	942	080	9.255	50
20	103	939	110	9.010	40
30	.1132	.9936	.1139	8.777	30
40	161	932	169	8.556	20
50	190	929	198	8.345	10
7°00′	.1219	.9925	.1228	8.144	**83°00′**
10	248	922	257	7.953	50
20	276	918	287	7.770	40
30	.1305	.9914	.1317	7.596	30
40	334	911	346	7.429	20
50	363	907	376	7.269	10
8°00′	.1392	.9903	.1405	7.115	**82°00′**
10	421	899	435	6.968	50
20	449	894	465	6.827	40
30	.1478	.9890	.1495	6.691	30
40	507	886	524	6.561	20
50	536	881	554	6.435	10
9°00′	.1564	.9877	.1584	6.314	**81°00′**
	Cos	Sin	Cot	Tan	DEGREES

Table of natural trigonometric functions (cont'd.)

DEGREES	Sin	Cos	Tan	Cot	
9°00′	.1564	.9877	.1584	6.314	**81°00′**
10	593	872	614	197	50
20	622	868	644	084	40
30	.1650	.9863	.1673	5.976	30
40	679	858	703	871	20
50	708	853	733	769	10
10°00′	.1736	.9848	.1763	5.671	**80°00′**
10	765	843	793	576	50
20	794	838	823	485	40
30	.1822	.9833	.1853	5.396	30
40	851	827	883	309	20
50	880	822	914	226	10
11°00′	.1908	.9816	.1944	5.145	**79°00′**
10	937	811	974	066	50
20	965	805	.2004	4.989	40
30	.1994	.9799	.2035	4.915	30
40	.2022	793	065	843	20
50	051	787	095	773	10
12°00′	.2079	.9781	.2126	4.705	**78°00′**
10	108	775	156	638	50
20	136	769	186	574	40
30	.2164	.9763	.2217	4.511	30
40	193	757	247	449	20
50	221	750	278	390	10
13°00′	.2250	.9744	.2309	4.331	**77°00′**
10	278	737	339	275	50
20	306	730	370	219	40
30	.2334	.9724	.2401	4.165	30
40	363	717	432	113	20
50	391	710	462	061	10
14°00′	.2419	.9703	.2493	4.011	**76°00′**
10	447	696	524	3.962	50
20	476	689	555	914	40
30	.2504	.9681	.2586	3.867	30
40	532	674	617	821	20
50	560	667	648	776	10
15°00′	.2588	.9659	.2679	3.732	**75°00′**
10	616	652	711	689	50
20	644	644	742	647	40
30	.2672	.9636	.2773	3.606	30
40	700	628	805	566	20
50	728	621	836	526	10
16°00′	.2756	.9613	.2867	3.487	**74°00′**
10	784	605	899	450	50
20	812	596	931	412	40
30	.2840	.9588	.2962	3.376	30
40	868	580	994	340	20
50	896	572	.3026	305	10
17°00′	.2924	.9563	.3057	3.271	**73°00′**
10	952	555	089	237	50
20	979	546	121	204	40
30	.3007	.9537	.3153	3.172	30
40	035	528	185	140	20
50	062	520	217	108	10
18°00′	.3090	.9511	.3249	3.078	**72°00′**
	Cos	Sin	Cot	Tan	DEGREES

Table of natural trigonometric functions (cont'd.)

Degrees	Sin	Cos	Tan	Cot	
18°00′	.3090	.9511	.3249	3.078	**72°00′**
10	118	502	281	047	50
20	145	492	314	018	40
30	.3173	.9483	.3346	2.989	30
40	201	474	378	960	20
50	228	465	411	932	10
19°00′	.3256	.9455	.3443	2.904	**71°00′**
10	283	446	476	877	50
20	311	436	508	850	40
30	.3338	.9426	.3541	2.824	30
40	365	417	574	798	20
50	393	407	607	773	10
20°00′	.3420	.9397	.3640	2.747	**70°00′**
10	448	387	673	723	50
20	475	377	706	699	40
30	.3502	.9367	.3739	2.675	30
40	529	356	772	651	20
50	557	346	805	628	10
21°00′	.3584	.9336	.3839	2.605	**69°00′**
10	611	325	872	583	50
20	638	315	906	560	40
30	.3665	.9304	.3939	2.539	30
40	692	293	973	517	20
50	719	283	.4006	496	10
22°00′	.3746	.9272	.4040	2.475	**68°00′**
10	773	261	074	455	50
20	800	250	108	434	40
30	.3827	.9239	.4142	2.414	30
40	854	228	176	394	20
50	881	216	210	375	10
23°00′	.3907	.9205	.4245	2.356	**67°00′**
10	934	194	279	337	50
20	961	182	314	318	40
30	.3987	.9171	.4348	2.300	30
40	.4014	159	383	282	20
50	041	147	417	264	10
24°00′	.4067	.9135	.4452	2.246	**66°00′**
10	094	124	487	229	50
20	120	112	522	211	40
30	.4147	.9100	.4557	2.194	30
40	173	088	592	177	20
50	200	075	628	161	10
25°00′	.4226	.9063	.4663	2.145	**65°00′**
10	253	051	699	128	50
20	279	038	734	112	40
30	.4305	.9026	.4770	2.097	30
40	331	013	806	081	20
50	358	001	841	066	10
26°00′	.4384	.8988	.4877	2.050	**64°00′**
10	410	975	913	035	50
20	436	962	950	020	40
30	.4462	.8949	.4986	2.006	30
40	488	936	.5022	1.991	20
50	514	923	059	977	10
27°00′	.4540	.8910	.5095	1.963	**63°00′**
	Cos	Sin	Cot	Tan	Degrees

Table of natural trigonometric functions (cont'd.)

DEGREES	Sin	Cos	Tan	Cot	
27°00′	.4540	.8910	.5095	1.963	**63°00′**
10	566	897	132	949	50
20	592	884	169	935	40
30	.4617	.8870	.5206	1.921	30
40	643	857	243	907	20
50	669	843	280	894	10
28°00′	.4695	.8829	.5317	1.881	**62°00′**
10	720	816	354	868	50
20	746	802	392	855	40
30	.4772	.8788	.5430	1.842	30
40	797	774	467	829	20
50	823	760	505	816	10
29°00′	.4848	.8746	.5543	1.804	**61°00′**
10	874	732	581	792	50
20	899	718	619	780	40
30	.4924	.8704	.5658	1.767	30
40	950	689	696	756	20
50	975	675	735	744	10
30°00′	.5000	.8660	.5774	1.732	**60°00′**
10	025	646	812	720	50
20	050	631	851	709	40
30	.5075	.8616	.5890	1.698	30
40	100	601	930	686	20
50	125	587	969	675	10
31°00′	.5150	.8572	.6009	1.664	**59°00′**
10	175	557	048	653	50
20	200	542	088	643	40
30	.5225	.8526	.6128	1.632	30
40	250	511	168	621	20
50	275	496	208	611	10
32°00′	.5299	.8480	.6249	1.600	**58°00′**
10	324	465	289	590	50
20	348	450	330	580	40
30	.5373	.8434	.6371	1.570	30
40	398	418	412	560	20
50	422	403	453	550	10
33°00′	.5446	.8387	.6494	1.540	**57°00′**
10	471	371	536	530	50
20	495	355	577	520	40
30	.5519	.8339	.6619	1.511	30
40	544	323	661	501	20
50	568	307	703	1.492	10
34°00′	.5592	.8290	.6745	1.483	**56°00′**
10	616	274	787	473	50
20	640	258	830	464	40
30	.5664	.8241	.6873	1.455	30
40	688	225	916	446	20
50	712	208	959	437	10
35°00′	.5736	.8192	.7002	1.428	**55°00′**
10	760	175	046	419	50
20	783	158	089	411	40
30	.5807	.8141	.7133	1.402	30
40	831	124	177	393	20
50	854	107	221	385	10
36°00′	.5878	.8090	.7265	1.376	**54°00′**
	Cos	Sin	Cot	Tan	DEGREES

Table of natural trigonometric functions (concluded)

Degrees	Sin	Cos	Tan	Cot	
36°00′	.5878	.8090	.7265	1.376	**54°00′**
10	901	073	310	368	50
20	925	056	355	360	40
30	.5948	.8039	.7400	1.351	30
40	972	021	445	343	20
50	995	004	490	335	10
37°00′	.6018	.7986	.7536	1.327	**53°00′**
10	041	969	581	319	50
20	065	951	627	311	40
30	.6088	.7934	.7673	1.303	30
40	111	916	720	295	20
50	134	898	766	288	10
38°00′	.6157	.7880	.7813	1.280	**52°00′**
10	180	862	860	272	50
20	202	844	907	265	40
30	.6225	.7826	.7954	1.257	30
40	248	808	.8002	250	20
50	271	790	050	242	10
39°00′	.6293	.7771	.8098	1.235	**51°00′**
10	316	753	146	228	50
20	338	735	195	220	40
30	.6361	.7716	.8243	1.213	30
40	383	698	292	206	20
50	406	679	342	199	10
40°00′	.6428	.7660	.8391	1.192	**50°00′**
10	450	642	441	185	50
20	472	623	491	178	40
30	.6494	.7604	.8541	1.171	30
40	517	585	591	164	20
50	539	566	642	157	10
41°00′	.6561	.7547	.8693	1.150	**49°00′**
10	583	528	744	144	50
20	604	509	796	137	40
30	.6626	.7490	.8847	1.130	30
40	648	470	899	124	20
50	670	451	952	117	10
42°00′	.6691	.7431	.9004	1.111	**48°00′**
10	713	412	057	104	50
20	734	392	110	098	40
30	.6756	.7373	.9163	1.091	30
40	777	353	217	085	20
50	799	333	271	079	10
43°00′	.6820	.7314	.9325	1.072	**47°00′**
10	841	294	380	066	50
20	862	274	435	060	40
30	.6884	.7254	.9490	1.054	30
40	905	234	545	048	20
50	926	214	601	042	10
44°00′	.6947	.7193	.9657	1.036	**46°00′**
10	967	173	713	030	50
20	988	153	770	024	40
30	.7009	.7133	.9827	1.018	30
40	030	112	884	012	20
50	050	092	942	006	10
45°00′	.7071	.7071	1.000	1.000	**45°00′**
	Cos	Sin	Cot	Tan	Degrees

INDEX

Index

DATE DUE

GAYLORD			PRINTED IN U.S.A.